CLASSICAL AND QUANTUM MECHANICS

NEW RESEARCH ON QUANTUM ENTANGLEMENT

CLASSICAL AND QUANTUM MECHANICS

Additional books in this series can be found on Nova's website under the Series tab.

Additional e-books in this series can be found on Nova's website under the e-books tab.

CLASSICAL AND QUANTUM MECHANICS

NEW RESEARCH ON QUANTUM ENTANGLEMENT

LORI WATSON
EDITOR

nova publishers
New York

Library of Congress Cataloging-in-Publication Data

New research on quantum entanglement / editor, Lori Watson.
pages cm -- (Classical and quantum mechanics)
Includes index.
ISBN 978-1-63482-888-8 (hardcover)
1. Quantum entanglement. I. Watson, Lori, editor. II. Series: Classical and quantum mechanics series.
QC174.17.E58N49 2015
530.12--dc23

2015016703

Published by Nova Science Publishers, Inc. † New York

CONTENTS

PREFACE

Quantum entanglement is a quantum mechanical phenomenon with applications in various fields. This book includes research on the time travel hypothesis, in the context of general relativity, the fundamental importance in the open quantum systems theory of the border line between "system" and "environment" and its applications; and a scheme of ghost imaging and the physics behind it with two EPR states.

After its first suggestion in the context of General Relativity, the time travel hypothesis is taking new strength starting from the post relativistic theories and particularly from the Open Quantum Relativity (OQR), which connects Quantum Mechanics and General Relativity, in a symplectic comprehensive picture. In Chapter 1, the existence of a five dimensional space and of two time arrows, come out as a natural necessity under the general assumption that the conservation laws can never be violated. The comparison between this dynamical theory and the experimental data in several fields of physics, is really satisfactory, so providing a solid check for the hypothesis of the two Time arrows existence and this existence can change dramatically any speculations about Time traveling. The OQR leads to some important consequences, like the existence of a dynamical unification scheme of all fundamental interactions, obtained by assuming a 5D space which allows that the conservation laws are always and absolutely valid as a natural necessity, while what we usually define as violations of conservation laws can be described by a process of topology change. Then a process of Embedding and dimensional reduction to 4D gives rise to an induced gravity matter theory, by which the usual masses, spins and charges of particles, naturally emerge. As we will see, the theory leads to a dynamical explanation of several problems of modern physics (e.g., entanglement of quantum states, mass generation, quantum teleportation, gamma ray bursts origin, black hole singularities, cosmic primary antimatter absence). The OQR provides also a self-consistent picture of the observed accelerated cosmological behavior, a correct calculation of experimental cosmological parameters and finally new predictions like gravitationally induced neutrino oscillations and additional scalar modes in gravitational waves. The theory shows the link between the geodesic structure and the field equations and, moreover, the emergence of an extra force term in the reduction process and the possibility to recover the masses of particles, allow to interpret the Equivalence Principle as a dynamical result which connect geodesics and metric structure. At the end, the Open Quantum Relativity is a covariant symplectic theory founded on a completely "affine" approach, following the Palatini formalism, which is more general with respect to the Levi-Civita connections imposed in General Relativity. In this new theory, the dimensional reduction gives rise to the generation of the masses of particles, which emerge from the field equations and the embedded geodesics and, in that way, the coincidence of chronological and geodesic structures is derived from the embedding and a new dynamical formulation of the Equivalence Principle is a consequence of dimensional reduction.

The OQR scheme is more general than the GR description, since two time arrows and closed time-like paths naturally emerge, and this fact is opening the way to further fundamental consequences, starting from a reinterpretation of the standard notion of causality which can be always recovered, even when it is questioned (like in entanglement and quantum teleportation) since it is enlarged and generalized to a "forward" and a "backward" causation. Finally, while the concept of time travels in GR is only related to the difference among distinct frameworks on one time arrow, here, in OQR, the relative difference can be also taken into account on two time arrows and this means that a closed timelike path does not need, to exist, special devices imposed "by hand", but is a feature of Nature, making a Time Machine conceivable.

Realistic many-particle systems dynamically exchange particles with their environments. In classical physics, small variations in the number of constituent particles are commonly considered practically irrelevant. However, in the quantum mechanical context, such and similar structural variations are generically taxed due to the so-called Entanglement Relativity. In Chapter 2 we point out difficulties in deriving master equation for a subsystem of an alternative partition of the closed quantum system. We find that the Nakajima-Zwanzig projection method cannot be straightforwardly used to solve the problem. The emerging tasks and prospects for the consistent foundations are examined.

A scheme of ghost imaging with two EPR states, *AB* and *CD*, is considered in Chapter 3. The scheme involves both the type-one and type-two ghost imaging arrangements. Photons *A* and *C* are coupled by a 50/50 non-polarizing beam splitter. Photon *A* passes through an object. Photons *B*, *C*, *D* go freely. Photon *D* is captured by a bucket detector. Photon *B* is captured by a detector, scannable along the transverse direction. A ghost image of the object is observed in the joint-detection between the detectors, with the maximum visibility 50%.

In: New Research on Quantum Entanglement ISBN: 978-1-63482-888-8
Editor: Lori Watson, pp. 1-98

Chapter 1

THE ENTANGLEMENT TIME MACHINE

***Giuseppe Basini*[1] *and Salvatore Capozziello*[2,*]**
[1]Laboratori Nazionali di Frascati, INFN,
Frascati, Italy,
[2]Dipartimento di Fisica, Universitá Federico II Napoli,
INFN, Sez. di Napolim Italy

Abstract

After its first suggestion in the context of General Relativity, the time travel hypothesis is taking new strength starting from the post relativistic theories and particularly from the Open Quantum Relativity (OQR), which connects Quantum Mechanics and General Relativity, in a symplectic comprehensive picture. In this framework, the existence of a five dimensional space and of two time arrows, come out as a natural necessity under the general assumption that the conservation laws can never be violated. The comparison between this dynamical theory and the experimental data in several fields of physics, is really satisfactory, so providing a solid check for the hypothesis of the two Time arrows existence and this existence can change dramatically any speculations about Time traveling. The OQR leads to some important consequences, like the existence of a dynamical unification scheme of all fundamental interactions, obtained by assuming a 5D space which allows that the conservation laws are always and absolutely valid as a natural necessity, while what we usually define as violations of conservation laws can be described by a process

*E-mail address: capozziello@na.infn.it (Corresponding author)

of topology change. Then a process of Embedding and dimensional reduction to 4D gives rise to an induced gravity matter theory, by which the usual masses, spins and charges of particles, naturally emerge. As we will see, the theory leads to a dynamical explanation of several problems of modern physics (e.g. entanglement of quantum states, mass generation, quantum teleportation, gamma ray bursts origin, black hole singularities, cosmic primary antimatter absence). The OQR provides also a self-consistent picture of the observed accelerated cosmological behavior, a correct calculation of experimental cosmological parameters and finally new predictions like gravitationally induced neutrino oscillations and additional scalar modes in gravitational waves. The theory shows the link between the geodesic structure and the field equations and, moreover, the emergence of an extra force term in the reduction process and the possibility to recover the masses of particles, allow to interpret the Equivalence Principle as a dynamical result which connect geodesics and metric structure. At the end, the Open Quantum Relativity is a covariant symplectic theory founded on a completely "affine" approach, following the Palatini formalism, which is more general with respect to the Levi-Civita connections imposed in General Relativity. In this new theory, the dimensional reduction gives rise to the generation of the masses of particles, which emerge from the field equations and the embedded geodesics and, in that way, the coincidence of chronological and geodesic structures is derived from the embedding and a new dynamical formulation of the Equivalence Principle is a consequence of dimensional reduction.

The OQR scheme is more general than the GR description, since two time arrows and closed time-like paths naturally emerge, and this fact is opening the way to further fundamental consequences, starting from a reinterpretation of the standard notion of causality which can be always recovered, even when it is questioned (like in entanglement and quantum teleportation) since it is enlarged and generalized to a "forward" and a "backward" causation. Finally, while the concept of time travels in GR is only related to the difference among distinct frameworks on one time arrow, here, in OQR, the relative difference can be also taken into account on two time arrows and this means that a closed timelike path does not need, to exist, special devices imposed "by hand", but is a feature of Nature, making a Time Machine conceivable.

1. Introduction. From Absolute to Relativistic Time

The definition of Time has been deeply modified along the last century and the concept of relative time, from the special Relativity up today, is not ceasing to change the physics, and moreover it seems that this revolution will produce even more consequences in the next future. The same concept of time machine, despite the difficulties that we have in conceiving it, comes out from the problem of the relative time deep meaning, which is still far to be fully settled. So than we should accomplish this passage and find a new synthesis of the scientific discoveries of last century, which are tesserae of a mosaic still incomplete. The traditional vision of time is no longer existent as "intrinsic" concept, and also the classical definition of absolute time, apparently reasuming this point of view, was in fact meaning a real change, because the innate concept of absolute Time becomes instead a physical definition. If the Time is considered a physical quantity, instead of a "pure fact", this involves a relation with physical laws, which can bring to a different definition, if these laws have to be changed. And this is what happened when Einstein made the revolution which, changing the formal relations between time and the framework of its definition, opened the way to all subsequent considerations on Time, including, as we will see, the matter of this paper. Already for Galileo and Descartes, time and space attain the role of topics of scientific investigation, in that way acquiring an operative function and becoming the ground on which the physical theories must be formulated. Some metaphysical remnants remain however in the absolute time and space, as defined by Newton in the Principia, because called: *"sensorium Dei"*. It is very interesting instead the position by Leibniz, which denies the absolute reality of space and time, defining them as relations among objects and phenomena. For Leibniz, time is the order of subsequent phenomena, while space is the order of coexistent objects and phenomena. In other words, in Leibniz, it seems present the implicit statement that without objects we have no space and time. This statement, really interesting, is in disagreement with respect to the Newton position, since, from his point of view, space and time are instead existing for themselves. As we know, the Newton viewpoint prevailed and so classical physics, all along 200 years, considered space and time as absolute concepts. The situation really changed only in the *XXth* century, recovering the Leibniz ideas with the advent of Special and General Relativity. In Einstein general equations, the time definition change dramatically, because is coming out from relations in which the light speed, in empty space, is the only existent absolute value, while time is not absolute, but only locally defined, depending on relative motions and masses of the considered frameworks. The concept of a "variable time" (together with the Galilean Experimental Method and the Uncertainty Principle) is really the most revolutionary concept of the history of physics, because it provided a completely new point of view and opened the doors to the following speculations on the Time nature, until the extreme hypotheses of time travels. Now, even the simple hypothesis of time travel can change dramatically our knowledge of the Universe, so than, even long before the OQR, many works and many scientist was devoted to this subject. Time travel has been considered for long time a dream, but Homer was saying that: *"Dreams come from Zeus"*.

Before presenting the OQR approach, we are going to resume the existent scientific positions on this intriguing problem, involving the time nature and the time travel possibility. Precisely, in section 2, we show how the problem of time travel comes out with the new phenomena like quantum teleportation, together with the ideas of scientists which are open to time travel possibility and people which instead do not believe at all in it; in section 3, we discuss the ideas of physicists which studied time travels in the rigorous limits of classical General Relativity; in section 4, we present the Open Quantum Relativity as intersection between Quantum Mechanics and General Relativity; in section 5, we define the Time travel possibility in this new Open Quantum Relativity framework; in section 6, time travels are discussed until the last consequences and in section 7, the conclusions.

2. Time Travel Hypothesis: The Classical Discussion

General Relativity is a new way to see the Nature and define many concepts, first of all Space and Time. In fact, following Special and General Relativity (always confirmed by a huge number of experimental observations) time is flowing slower in a framework in motion (or close to a larger mass) with respect to an observer at rest, allowing the extreme hypothesis (revolutionary, even if still with strong limitations) of time travels, (*e.g.*, like in the famous twins paradox, where one of the brothers, travelling in a faster rocket for a significant time, coming back home, would find his twin evidently older than him). After that, the debate started up if travelling in time would be confined in the limits of classical interpretation of General Relativity or also the possibility to travel in the past would have been possible. This debate takes a new boost after the introduction, by Goedel, of the Closed Timelike Curves (CTCs), [16, 42] which, including the concept of a circular time trajectory, would make intrinsically possible "to pass again" in the past. The Goedel idea of considering cylindrical space time coordinates, where space is described by longitudinal coordinate and time by the circular one, instead of the classical contrary, even if only of mathematical (not dynamical) origin, was able to give a solution of Einstein equations. And in fact, after Goedel, many authors showed that the possible existence of natural CTCs, in the Universe, should be not an exception, but an infinity, unless the existence of some ad hoc rules could avoid it. If the possibility of "relative" time travels in the future (*e.g.*, for an astronaut in the future of the human kind on Earth, if accelerated for a significant period in a rocket outside the Earth itself) is possible already in the General Relativity (section 3) it is in the post relativistic theories, including also Quantum Mechanical effects, that the time travel hypothesis takes a more general meaning (section 5). The time travel hypothesis is the first of the open problems, because concerning physics and logic together. In any case, what we should stress is that: a) if we look at the shortcomings of modern physics (like singularities or logic conflicts, as for black holes or the entanglement phenomenon), the Time concept plays a unique role and the new crucial question comes out because deeply related to it, b) the non-existence of a quantum relativistic theory, made, up to now, very difficult to match results and predictions of Quantum Mechanics with those of General Relativity.

The consequence was the growing interest about post-relativistic and quantum field theories (like *e*.g. the Many Worlds Theory[1], the Superstrings [2] or the Supergravity [3], but very probably only a full quantum field relativistic theory can really settle the problem of Time and Time travels. In this work, after a recall of the attempts, conventional and unconventional, made in last decades, we are going to show how the new Open Quantum Relativity (OQR) theory can frame Time and Time Travels, including the possible explanations of the new and very surprising experimental results.

2.1. The Chronology Protection Conjecture

Researches involving the entire scientific community started up after the General Relativity, authors like Goedel, Wheeler, Penrose, Amaldi, Guth, Schmidt, De Witt and many others, made a lot of effort in speculating and searching on the Relativity and the Time concept. The first and more traditional school that we encounter, on this topic, is that one of people excluding time travel possibility. The Chronology Protection Conjecture (name introduced by Hawking [4] is in fact a class of conjectures elaborated to take into account the fact, considered as established, that travels in time seems up to now unobserved, even if several theories are not excluding it. "Why, if time travels was allowed, we are not full of time voyagers?" This sentence comes from Steven Hawking himself, who invented a mechanism able in principle to avoid the occurrence of CTCs. In fact, his conjecture [4] is more devoted to "localized" CTCs, because he refused any general hypothesis of motion (for instance in a wormhole) in the "imaginary direction" of Time. Starting from a criticism on Goedel type solutions (seen as pure formal solutions) Hawking then explored the possibility of localized CTCs, confined in chronology violating regions (where the causality seems violated) contained in a chronological horizon and it is the behaviour around this horizon, which provides the basis for the Chronology Protection. The chronological horizon is defined as the boundary of the future for the chronology violating region and is a Cauchy-type horizon and Hawking considered that, as a time machine (which, in this sense, means a region with CTCs) is formed the first null and timelike curves appear. Even not rejecting completely the idea that "if there is a timelike tube T connecting surfaces S and S' of different topology, then the region $\mathcal{M}t$ contains CTCs", for Hawking no matter or informations, can cross the chronological horizon into the region of CTCs and moreover if no CTC is present initially, one cannot create them. M. Visser at conference held in Cambridge in honour of Hawking himself [5] summarized it with a good picture. Under acceptable conditions, a chronology horizons, appropriate to a locally constructed time machine, should be compactly generated and contain a sort of "fountain" (in essence the first closed null curves to come to existence as the time machine is formed). Here a classical photon will circulate around the fountain infinitely many times, in effectively zero elapsed time and, on each circuit, there is a non trivial holonomy changing the energy of the photon. For a past chronology horizon, which expands as we move to the future (as defined by an observer outside the chronology violating region), this provides a net increase in the photon energy for each circuit of the fountain. The photon energy geometrically increases reaching infinity in effectively zero time. On each circuit:

$$E \to e^{h}E \to e^{2h}E...., \quad h = -2\oint \mathcal{R}\,(\epsilon)dt, \tag{1}$$

with the size of the energy boost being controlled by a loop integration around the fountain involving the Newman-Penrose parameter ϵ. The source of this energy must ultimately be the spacetime geometry responsible for the chronology horizon and, by extension, the stress-energy used to warp space-time and set up the fountain. If we now let the photon (and the gravitational field which it generates) back-react on the spacetime, its infinite energy will presumably alter the space time geometry beyond all recognition. This is in summary the Hawking Chronology Protection Conjecture, which he derives from classical arguments (a classical point particle following a defined null curve), but now, what happens if we take into account quantum effects, the conjecture is still working or not? Opinions was divided on that, and still they are.

2.2. Time Travels. In Favour and Against the Chronology Protection

In the rigorous limits of classical General Relativity, another claim of impossibility "to build" a Time machine (i.e., to warp, by human devices, the spacetime up to CTCs appears) comes by Krasnikov [6]. He, even not excluding the natural appearance of CTCs, presents a theorem which states: "Any space time U has a maximal extension *Mmax* such that all closed causal curves in *Mmax* (if they exist) are confined to the chronological past of U". For Krasnikov, this means that a time machine cannot be built. But what kind of time machine? Only a time machine like the one by Gott [7] or by Morris, Thorne and Yourtsever [8], that is a General Relativity time machine (even if involving cosmic strings) and this is a real limit, because it is not sure at all that classical General Relativity, not including Quantum Mechanics, can be considered a good base for such considerations.

In particular, authors like Deser, Jackiw and t'Hooft [9] states that physical cosmic strings cannot generate CTCs. They reexamine the causal properties of geometries generated by parallel, moving cosmic strings and, contrary to other claims, they say that acausal behaviors cannot be realized by physical, timelike sources. (anyway Krasnikov himself says that classical General Relativity does not fully determine the evolution of the Universe, so then, when a theory solving this problem will emerge, it may happen that the time machine creation would be possible). Another supporter of the chronology protection is W.H. Huang, which shows that the renormalized stress-energy tensor may be divergent, at the chronology horizon. Huang first makes a critical revue of the arguments of some opponents of chronology protection, like for instance Boulware [11] or Tanaka *et al.*, [12] which, for a sufficiently massive field, claim for the regularity of the stress-energy tensor at the chronology horizon in Gott space and Grant space, or Lyutikov [12] which, making a quest of the Roman ring of traversable wormhole, find that the vacuum fluctuation can be made arbitrarily small. Then

Huang presents a calculation for which at least one component of the renormalized stress-energy tensor must always be divergent at the chronological horizon. Using his words: "Our investigation suggests that the physical law of requiring the finite stress-energy tensor will forbid a geometry with the causality violation to be enclosed in a region which maintains the required condition of global causality". Finally Cassidy [14] supported the Cronology Conjecture remembering that, even if a self-consistent vacuum state in the Misner spacetime could be constructed, it must be unstable against a slight perturbation. Against the Hawking chronology protection is instead Gonzalez-Diaz [15], which points out that the main problem of the time machines, is the emergence of divergences of the stress-energy tensor of the quantum vacuum fluctuations on the chronology horizon. This final point is in fact the major fundament of the chronological conjecture, because it would make unstable the space time constructed so preventing time travel to occur.

But Gonzalez-Diaz objected that this conjecture is based on semi-classical considerations, so then no longer holding in the framework of quantum gravity, because spacetime foam must entail strong violations of causality everywhere and so "the divergences of stress energy tensor due to quantum fluctuations merely are artifacts, from inappropriate combination of quantum matter field with a classical spacetime device". Also Visser [5] criticized the chronological protection conjecture, because based on classical relativity arguments, and, once taken into account quantum effects, immediately arises the question if those effects destroy the conjecture since playing a dominant and opposite role. He says: "Real photons are wave packets with a certain transverse size, and generically the same effects that leads to the energy being boosted (so producing the instability) leads to the wave packet being defocused - the geometry in the surrounding tube-like region acts as diverging lens. With two competing effects which one wins?" In fact there are geometries for which the defocusing effect overwhelms the boost effect and the stress-energy tensor remains finite, but also other geometries for which the reverse holds true and this means that classical effects do not provide a definitive answer. (For example the search for Lorentian Traversable Stable Wormholes is going on). Stable Wormholes model are presented by Kanty *et al.*, in 2011 [64] and by Novikov and Shatskiy in 2012 [64] where is demonstrated the wormhole stability under spherical perturbations. Against the chronological protection is also Li Xin Li [16] which indicates that the appearance of absorption material, in the spacetime containing CTCs, can cause that the renormalized stress-energy tensor of vacuum fluctuations should be convergent at the Cauchy horizon (the chronological horizon dividing regions of normal causal behaviour from regions of abnormal causal behaviour, i.e., violating the chronology) and such a spacetime might be stable. Considering any sort of high energy scattering (in the extreme relativistic limit) Li Xin Li stressed that, for many processes, the total cross section σt increases with the incident energy or tends to a constant. It turns out that it seems that materials will become opaque at extremely high energy of incident particles or waves and the cross section for gravitational scattering for two identical particles σ_t is

$$\sigma_t \sim G^2(\hbar\omega)^2/c^8 \quad \hbar \gg mc^2 \tag{2}$$

where σ_t might act as an universal cross section for absorption caused by scattering of all kinds of incident waves, including the scalar, the spinor, the vector and the tensor waves, since all particles are subjected to the gravitation. Following the total cross section behavior, one can divide reactions in three classes: σ_t increases with ω through

$$\sigma_t = \alpha\omega^s, \quad s > 0, \quad \omega \gg \omega_0$$

or

$$\sigma_t = sln(\omega/\omega_0) \quad s > 0, \ \omega \gg \omega_0$$

(where ω_0 is the critical values of ω , and α is the fine structure constant). Finally, σ_t tends to a constant

$$\sigma_t \to \pi r_0^2, \ as \ \omega \to \infty$$

Li Xin Li shows (through the Hadamard functions) that, for all three cases, it is always possible to find conditions for which the renormalized stress energy tensor of the vacuum fluctuations does not diverges at the Cauchy horizon. For Li Xin Li, if absorption material with appropriate density is introduced, all vacuum fluctuations can be smoothed out, in such a way that the metric perturbation will be very small and the spacetime with time machine may be stable. The conclusion of the author is very radical: "Contrary to the chronology protection conjecture, I give the anti-chronology protection conjecture: there is no law of physics preventing the appearance of CTCs". Also inside the narrow limits of General Relativity the debate is fully open, even if, in fact, it is outside those limits that it takes more strength.

2.3. New Questions: Quantum Effects, Teleportation and Tachyons

A line of research which is very interesting is the so-called quantum teleportation. In fact, after the long period occurred between the EPR paradox formulation and the Bell theorem [17, 18], the question of the entanglement was reopened, when experimental e vidences came out demonstrating that Quantum Mechanics is correct even if contradicting the locality principle. Beside the pioneeristic works of Aspect [19], many other experiments were performed indicating that, under some circumstances, interaction between objects without any apparent possibility of mutual influence (e.g., to be connected in the light speed limit) confirmed the validity of the Schröedinger hypothesis of an "entanglement" as a feature underlying the usual measurable properties. After that, a long series of works investigated the characteristics of the new phenomenon and tested the possibilities to use it to transfer information instantaneous and impossible to be detected outside the entangled emitter-receiver system. The long series of Bob & Alice (receiver & emitter) papers is the most famous example of it [20]. Very important was the claim by Bouwmeester *et al.*, [21], where they presented experimental evidences of an effect of polarization, given on one of two entangled photons, transferred on the other one without any direct standard interaction between them. For Bennett *et al.*, [22]: "It is possible to transfer the quantum state of a particle onto another particle, provided one does not get any information about the state in the course of the transformation". Bouwmeester take as starting point (Alice) a single photon 1 which can be polarized vertically $|\updownarrow\rangle_1$ horizontally $|\leftrightarrow\rangle_1$ or defined by a general superposition of these two states

$$| \psi\rangle_1 = \alpha\, |\leftrightarrow\rangle_1 + \beta\, |\updownarrow\rangle_1, \tag{3}$$

where α and β are two complex numbers such that $|\alpha|^2 + |\beta|^2 = 1$. After passing in a polarizing beam splitter, the photon 1 will be reflected (or transmitted) with probability $|\alpha|^2$(or $|\beta|^2$) and the general state $| \psi\rangle$ would be projected either onto $|\updownarrow\rangle$, or onto $|\leftrightarrow\rangle$, by the action itself of the measurement. So, following Quantum Mechanics, Alice cannot reconstruct the state of photon 1 but, however, she can send it to Bob by using another pair of entangled photons, photon 2 and photon 3, shared with Bob, entangled in the state

$$| \psi^-\rangle_{23} = \frac{1}{\sqrt{2}}(| \leftrightarrow\rangle_2 | \updownarrow\rangle_3 - | \updownarrow\rangle_2 \; | \leftrightarrow\rangle_3). \tag{4}$$

The teleportation works in this way: Alice has photon 1 in the initial state and also photon 2, entangled with photon 3 which is in the hands of Bob. The essential point is to perform a specific measurement on particles 1 and 2 able to project them onto a new entangled state

$$|\,\psi^{-}\rangle_{12} = \frac{1}{\sqrt{2}}(|\leftrightarrow\rangle_1|\updownarrow\rangle_2 - |\updownarrow\rangle_1|\leftrightarrow\rangle_2). \qquad (5)$$

(Remember that this is only one of the four possible entangled states, into which a state of two particles can be decomposed, but it distinguishes by the fact that it changes sign upon interchanging particles 1 and 2, and this antisymmetric feature, utilized in the experimental identification, is an information which has to be sent to Bob in the standard way).

Quantum physics [22] predicts that once particles 1 and 2 are projected into $|\psi^{-}\rangle_{12}$ state, particle 3 (previously entangled with 2) is instantaneously projected into the initial state of photon 1, due to the fact that now, whatever the state 1 is, photon 2 must be in the opposite state (orthogonal) of 1. But photon 2 and photon 3, at the beginning, were prepared in the state $|\psi^{-}\rangle_{23}$ (3 was also orthogonal), so the only possibility is that now photon 3 is in the same state as was photon 1, that is

$$|\psi\rangle_3 = \alpha|\leftrightarrow\rangle_3 + \beta|\updownarrow\rangle_3 \qquad (6)$$

This at the end is an "entanglement swapping", photon 1 is no longer available in the original state, but photon 3 is now in that state and this is really the result of teleportation. It remains that it is necessary to send to Bob, in standard way, the information of the state of the entanglement (one possibility among four) between photon 1 and photon 2, and this opened a serious discussion on the real meaning of teleportation, because: a) it seems possible to determine the entanglement also between particles (and very probably also between more complicated objects) without the same origin, so making conceivable a general technique; b) it seems possible to send information, via entanglement, without any destroying influence of the environment and as far as one wants; c) it seems possible to send information instantaneously, even if a conventional message is necessary to inform and check.

The last two points seem to indicate a $\Delta t = 0$ in the transfer operation, the only one way to send information instantaneously as far as we want. This really seems in contradiction with Special Relativity, even if some authors are saying that, due to the necessity to send also a conventional message, this would restrict again the phenomenon in the limits of Relativity [23]. This interpretation is really not fully convincing, not only because it is not completely sure that the conventional message is an intrinsic and permanent necessity, but also because, in any case, the entangled information remains instantaneous and not influenced at all during the travel. The comprehensible tendency to search a simple composition between the new results and the Relativity can be at the end misleading. What we want stress is that there are at least two problems, namely, if this is whether or not in contradiction with Relativity and how the mechanism works (i.e., what is its real dynamics). In fact, the entanglement, in the traditional picture, was not a phenomenon based on a known dynamics and, moreover, an instantaneous action (whatever its nature is) remains not compatible with the finite value of light speed. Quantum teleportation, is one of the most promising and, at the same time, problematic question of modern physics.

Going on in Time-travel physics, also when out of the mainstream (and this is the case), we will now deal with tachyons, particles supposed faster than light, whose existence would imply time travels. First of all, we have to say that tachyons could be solutions of every field theory where the so called "on mass shell condition" $p^{\mu}p_{\mu} = m_0^2$ is violated, in the sense that we can have

$$p^{\mu}p_{\mu} = -m_0^2 \tag{7}$$

with $c = 1$. The possible interpretation of that is twofold: a) the orthodox approach (absolute validity of Relativity postulates, with c maximal limit of every speed) which rules out such solutions as "hunphysical", since the masses of particles are imaginary; b) the alternative view that particles such that $\nu > c$, can exist and so retain such solutions. In fact, tachyons are taken into account in some interpretations of several extended unification schemes, as String Theory [2], Supergravity [3] or Brane Theory [24], where they are considered as extra-degrees of freedom, which could dynamically furnish solutions of problems as the compactification or the dark matter. For instance, tachyons arise in string theory whenever there is no space-time supersymmetry. In the bosonic version of string theory, all branes have tachyons, while in Superstring Theory they appear as modes on branes or as brane-antibranes pairs.

Practically, it is taken in consideration a tachyon field T and its dynamics is studied starting from an action of the type:

$$A = -\int d^{n+1}xV(T)\sqrt{-det(g_{\mu\nu} + \partial_{\mu}T\partial_{\nu}T)}\ , \tag{8}$$

where $g_{\mu\nu}$ is the metric and n the number of spatial dimensions, while $V(T)$ is the tachyon potential, characterized by an unstable maximum at $T = 0$ and a minimum at T = ∞.

From these quantum field theory viewpoints, tachyons can be considered under the standard of tools capable of solving dynamical problems (e.g., a rolling tachyon, in string cosmology, can lead the dynamics of dilaton giving rise to inflationary solutions [26].) However, in this work, we are interested into the fact that their existence implies a causality violation, because tachyons can travel faster than light. Due to this fact, the role of an emitting source of tachyons and of a receiving detector is inverted, so that the effect could come before the cause [25]. This statement can be easily demonstrated considering standard Lorentz transformations which relate two space-time events (e.g., a source and an observer). Such events are separated by time and space intervals $\Delta t > 0$ and $\Delta x > 0$. In a coordinate frame $\{t', x'\}$ moving with relative speed $0 < u < c$, the event are separated in time by

$$\Delta t' = \frac{\Delta t(1 - uv/c^2)}{\sqrt{1 - u^2/c^2}} \tag{9}$$

where $\nu = \Delta x/\Delta t$. Now we take into account tachyons with $\nu > c$ which are, respectively, emitted and absorbed in the two space-time events. The relative speed of the two frames is u which can be $c > u > c^2/\nu$. In this case Δt and $\Delta t'$ have opposite signs and then the time order of these events is interchanged in the two frames : what is emission in the first frame is absorption in the second one and vice-versa. Similar result if we take into account energy in the two frames. We have

$$E' = \frac{E - up}{\sqrt{1 - u^2/c^2}} \tag{10}$$

with the energy and the momentum defined as

$$E = \frac{m_0 c^2}{\sqrt{\epsilon(1 - v^2/c^2)}}; \quad p = \frac{m_0 v}{\sqrt{\epsilon(1 - v^2/c^2)}}. \tag{11}$$

The constant ϵ is -1 for tachyons and 1 for standard particles. The quantity E' can be positively or negatively defined, depending on the frame, while it is always positively defined in standard Relativity. It is interesting to see that the existence of tachyons allows time travel (i.e., toward the past) but this fact is conflicting with the classical notion of causality which, in this case, should be revised. This consideration imply a deep critic to the postulates of Special and General Relativity. The weaker point of this hypothesis is that it is not clear if tachyons are conceived as detectable or undetectable by definition, because even if their "hunphysical region" can be postulated existent, maybe they can never be detected in the region where we, the observers, live, so then they are out of the experimental physics in the Galilean sense. In every way, tachyons are at the end highly questionable and, since now, no convincing test has been furnished to demonstrate their existence. In conclusion, with the present state of knowledge, a tachyon time-machine seems very far to be really conceivable.

3. Time Travels in General Relativity

First of all, let us deal with time travels in the framework of General Relativity, showing that the topological and causal structure of this theory, does not prevent with certainty such a possibility, allowing CTCs. If definitely demonstrated, this possibility will be a major change, but beside that, new assumptions are, in any case, probably necessary, because the whole scenario is so changed by the last theoretical evolution and experimental data, that there are problems that have to be settled in a necessary new manner. Before examining new perspectives in post-relativistic theories, we review the current models and theories in the rigorous limit of standard General Relativity.

3.1. Causal and Topological Structure of Spacetime in General Relativity

In General Relativity it results very important the search of the possible existence of spacetimes with non-trivial topological structure. For example, the wormhole solutions of Einstein equations described by Wheeler [27, 28] are well-known examples of spaces with such a topology. The simplest example of this kind of spacetime is the Einstein-Rosen bridge [29] where the spacetime of an eternal black hole can be interpreted as an evolution reducing the bridge to a singularity, because, following the time arrow toward the future, the throat shrinks to zero size and the singularity arises. No causal signal can propagate through the throat, from one asymptotically flat region $(R+)$ to another one $(R'+)$, property which is connected with the particular causal structure of spacetime, in the presence of a black hole.

Three-dimensional geometries, with a wormhole connecting different asymptotically flat regions of the same spacetime, can be regarded to argue the initial value for vacuum solutions of the Einstein equations. In absence of matter, the throat of the wormhole pinches of so quickly that it cannot be traversed even by light, and this is a feature of spacetimes with a non simply connected Cauchy surface. Let's recall the theorem demonstrated by Gannon [30] (see also Lee [31], Galloway [32]): *any asymptotically flat spacetime, with a non simply connected Cauchy surface, has singular time evolution if it satisfies the weak energy condition.* Following this theorem, we can say that such singularities arise so quickly, that no external information signals can propagate through a wormhole during its life, up to the asymptotic region before the creation of the singularity. It seems to exist a sort of topological censorship, which says that no observer remaining outside a region with strong gravitational field, has time to probe the topology of spacetime [33]. In order to prevent shrinking of a wormhole and to make it traversable, one needs to fill its throat with matter, violating the averaged null energy condition. On the existence of matter violating the weak energy condition, several papers have been published, see e.g., Thorne [34], Visser [35], Flanagan and Wald [36]. A spacetime with a traversable wormhole can be transformed into a spacetime with CTCs, as it was shown by Morris, Thorne, Yurtsever [8] and Novikov [37] which showed that CTCs may arise as a result of the relative motion of the wormhole's mouths. The possibility to create CTCs in a space time where they were before absent, generated great interest toward the wormhole-like solutions and this result hints at deep relations between the topological and causal structure of space time. This implies that if such a space time is a solution of the Einstein equations, there must be a delta-like distribution of stress-energy at the junction between the two mouths, which violates the null energy condition, because any bundle of radially travelling null geodesics, that passes through the wormhole, is converging as it enters and diverging as it leaves, and therefore gets defocused by the wormhole. Whereas the mouths' intrinsic geometries are the same, the proper time interval between two identified events on the mouths must be the same as seen through either mouth. If the mouth is set in motion in the external space, assumed that the position of the other mouth stay unchanged, one can decide the clocks located near both mouths to be synchronized in the external space before the beginning of the motion. As a result of the motion, the synchronization in the external space is lost. The moving clock (measuring proper time near one mouth) shows less time than the clock which remains at rest. Denoted by $\Delta\tau$ this time difference, if $\Delta\tau > L/c$ where L is the distance between the final positions of the mouths (depending on the path), a CTC becomes possible. Denoted by C the region formed by events through which a CTC passes, the region formed by adding to C events which can be connected with those in C by causal past directed lines, is restricted from the past by the future chronology horizon.

In the external space, there is a dilation of proper time on the moving mouth relative to the static one, but, as seen, through the wormhole there is no such time dilation. CTCs could also arise [8] in a simpler model, where the mouths are moving with constant velocity with respect to each other and, in this case, CTCs are confined to a so called bounded achronal region of spacetime: the region that begins at the future chronology horizon and ends at the past chronology horizon (defined similarly to the future chronology horizon with the obvious change of the direction around the CTCs which enter the definition).

Achronal regions containing CTCs is a general property of space times with wormholes, because generic relative motions of wormhole's mouths will always produce CTCs [8] as will produce the gravitational redshifts when the wormhole's mouths are placed in a generic gravitational field. Let's now put in relation the conditions of CTCs formation with topological invariants of the multiply connected spacetime [39]. This overview is important to recognize the structure of CTCs and what are the conditions which allow time travels in classical General Relativity. We show at first that the gravitational field, in a locally static multiply connected spacetime, is generically non-potential. A spacetime is called locally static if, in any simply connected region, there exists a uniquely defined timelike covariant Killing vector ξ obeying the relations

$$\xi_{(\mu;\nu)} = 0, \quad \xi_{[\mu;\nu}\xi_{\lambda]} = 0 \tag{12}$$

The first equation is the definition of a Killing vector, while the second means that ξ is locally "surface orthogonal". In a multiply connected spacetime (i.e., when a wormhole is present), it is impossible to define such a Killing vector globally. The presence of non-contractible closed paths, which pass through a wormhole and cannot be contracted to a point by continuous transformations, may create an obstacle for the existence of a global Killing vector field, because, after continuous propagation of a Killing vector field along a closed non-contractible path, it will return to the initial point with the same direction as the initial vector, but with a different norm. In order overcome this situation, it is useful to introduce a unit timelike vector $u^\mu \equiv \xi^\mu/|\xi^2|^{1/2}$ instead of ξ. The four-velocity of a Killing observer as well as his four-acceleration $w^\mu \equiv u^\nu u^\mu_{;\nu}$ do not depend on the norm of ξ, and hence they are well defined globally. It is easy to verify that the following equations

$$u^\mu u_\mu = 1, \quad u_{\mu;\nu} = w_\mu u_\nu, w_{[\mu;\nu]} = 0 \tag{13}$$

are locally equivalent to Eqs. (12). So then, a locally static spacetime can be equivalently defined as a spacetime which admits two globally defined vector fields u^μ and w^μ obeying the relations (13). We can consider a locally static spacetime M as the collection S of Killing trajectories [40], *i.e.*, the set of integral lines $x^\mu(\tau)$ of u^μ being $dx^\mu/d\tau = u^\mu$. We assume that a unique Killing trajectory passes through each point of the spacetime $\mathcal{M}$ and that $\mathcal{M}$ = $\mathcal{T}$ x $\mathcal{S}$ where $\mathcal{T}$ topologically is the axis of real numbers R^1, and S the three-dimensional space under consideration. A non-contractible loop, passing through a wormhole n-times in the chosen direction, can be characterized by its winding number n and two loops with the same winding number n are connected by a continuous transformation (they are *homotopic*) and these loops are also homological, i.e., they are boundaries of some two-dimensional surface. In other words, the first fundamental group π_1 and the first homology group H_1, calculated for $\mathcal{S}$ (as well as for the spacetime $\mathcal{M}$ itself) coincide with the Z-group of integer numbers. The last equation in (13) shows that the acceleration form $w = w_\mu dx^\mu$ is closed ($dw = 0$). Thanks to the Stoke's theorem, the integral of this form over any closed path Cn is

$$I_n[w] = \oint_{Cn} w \tag{14}$$

and depends only on the winding number n of the path, with $I_n = nI$. We select the direction of the path with $n = 1$ in a way that $I[w] \equiv I_1[w] \leq 0$. If the period I $[w]$ vanishes, then the form w is exact, $i.e.$ there exists a globally defined scalar function φ (gravitational potential) such that $w = d\varphi$. For such a *potential gravitational field* the vector field $\xi^\mu = e^\varphi u^\mu$ is a global Killing vector field. In the general case (when $I[w] \neq 0$), the gravitational potential φ cannot be defined as a single-valued function. Such a gravitational field is so called *non-potential* and the period $I[w]$ may be considered as the measure of non-potentiality of the field. The interesting physical feature, of a non-potential gravitational field, is that the work done by such a field, on a particle moving along a closed contour Cn passing through a wormhole, does not vanish. This work is proportional to $e^{ln[w]}$. It is relevant now to see the characteristics of spacetime with a non-potential gravitational field, but, before, let's have a look on *clock synchronization*.

3.2. The Problem of the Synchronization

The *clock synchronization*, in a multiply connected spacetime, is a general problem, and the presence of non contractible paths makes this problem not neglibible also in a potential gravitational field. In fact, in a static gravitational field, it is possible to synchronize the clocks along any path lying in a simply connected region, but for the synchronization along the closed path passing through a wormhole, one cannot guarantee that the "synchronized clocks" show the same time as the initial ones. Generally, there will be a time gap for the clock's synchronization along a non-contractible path. For a potential gravitational field, the value of the time gap for clock synchronization, in a multiply connected spacetime, is a topological invariant. In order to describe this, one must introduce the 1-form:

$$\eta \equiv e^{-\varphi} u_\mu dx^\mu \tag{15}$$

It is possible verify that this globally defined 1-form is closed $(d\eta = 0)$. In fact, using equations (13) and the definition of the acceleration, $w_\mu = \varphi_{,\mu}$ we have

$$\eta_{[\mu;\nu]} = e^{-\varphi}[u_{[\mu;\nu]} - \varphi_{[,\nu}u_{\mu]}] = e^{-\varphi}[w_{[\nu}u_{\mu]} - \varphi_{[,\nu}u_{\mu]}] = 0 \tag{16}$$

The integral of the closed 1-form η over a closed path C_n

$$J_n[\eta] = \oint_{C_n} \eta, \tag{17}$$

is a topological invariant depending only on the winding number.

The value of $J_n[\eta]$ coincides with the time gap, for clock synchronization along a closed path with a chosen winding number n. In fact, chosen a point p_0 and labelled by γ_0 a Killing trajectory passing through this point, we take into account a curve $\sigma_\Gamma[p_0]$, which begins at p_0, passes once through the wormhole, and ends at the point p'_0, lying on the same Killing trajectory as p_0. We consider that the curve $\sigma_\Gamma[p_0]$ is every where orthogonal to the four-velocity of a Killing observer and this condition means that this curve is formed by the events which are simultaneous (in the reference frame of Killing observers) with the initial event p_0. The projection Γ of this line into the space S is closed, but in the general case the line itself is not closed, in other words, we can change the projection to get the evidence that the line is closed. We apply the component of a Killing trajectory lying between p_0 and p'_0, to this curve, to make it closed. Evaluating the integral (14) for this closed path with a winding number n = 1, is relatively simple and we can see that only the part of the path lying along the Killing trajectory gives a non-vanishing contribution. At the end, the value of the integral $J_1[\eta]$ calculated for the chosen path coincides with the time gap, for the clock synchronization along this path. It is evident that does not depend on the particular choice of path with the given winding number $n = 1$ and that $J_n[\eta] = nJ_1[\eta]$. In a non-potential gravitational field, the time gap $\Delta_1(\tau_0) \equiv \tau_0 - \tau_1$ for clock synchronization is time dependent. It can be shown [39] that it satisfies the equation

$$\frac{d\Delta_1(\tau_0)}{d\tau_0} = 1 - e^{I[w]} \tag{18}$$

This relation shows that the time gap for clock synchronization, along any closed path with a winding number $n = 1$, grows with time and the value $I[w]$ is the measure of the growth. When the gap becomes greater than the minimal time, necessary for a light signal to propagate along a closed path passing through the wormhole and return, a CTC can arise. The interaction of a wormhole with classical matter generically generates a non-potential component of the gravitational field (Frolov and Novikov [39]), because a locally static wormhole is generically unstable with respect to the processes which could transform it into a "time machine". As seen [8], relative motion of the wormhole's mouths also generates a time gap for clock synchronization We can interpret these results saying that there exist inner relations between the topological and causal properties of a spacetime, and, at the end, the existence of CTCs is a general feature of multiply connected locally static spacetimes [35].

3.3. Spacetimes with Closed Timelike Curves and the Issue of Time Machine

It is long time that we know solutions of Einstein's equations allowing CTCs, and the oldest example of a spacetime with CTCs (see also [34, 35, 100, 4]) is a solution obtained by Van Stockum in 1937 [41], which describes an infinitely long cylinder of rigidly and rapidly rotating dust. Another fundamental example and perhaps the most famous, is Goedel solution [42] representing a stationary homogeneous universe with non zero cosmological constant, filled with rotating dust. CTCs are also present in the interior of Kerr black hole in the vicinity of its ring singularity and other examples were discussed by de Felice [43]. Generally speaking, a spacetime can be divided into chronal regions, without CTCs, and achronal regions whit CTCs. The boundaries between the chronal and achronal regions are formed by chronology horizons, namely chronal regions end and achronal region begins at a future chronology horizon, while achronal regions end and chronal region begins at past chronology horizon. So then, achronal regions are intersections of the regions bounded by both of these horizons. A future chronology horizon is a particular type of Cauchy horizon, and it is subject to all the properties of such horizons. More precisely, it is generated by null geodesics, that have no past endpoints, but can leave the horizon when followed into the future. The spacetime future chronology horizon is defined as compactly generated if the generators, monitored into the past, enter one or more compact regions of spacetime and never thereafter leave them, as an example in a wormhole model with CTCs the future chronology horizon is compactly generated. A compactly generated chronology horizon cannot form in a spacetime derived from a spacelike non-compact surface without boundary, if the null energy condition holds [4]. The inner horizon of a Kerr-Newman solution is an example of a Cauchy horizon that is not compactly generated. In the context of General Relativity, in order to create a time machine (which, in this sense, means in general a region with CTCs) by using a wormhole, one needs to assume that there exists the possibility, in principle, to make them long living and traversable. But this is impossible without violation of the averaged null energy condition. This means that it is impossible to create a time machine in a finite region of spacetime without violating the averaged null energy condition. It is still not clear whether it is possible to provide such a violation in physically acceptable conditions, even if it has been shown that the averaged null energy condition is satisfied, for non interacting quantized scalar and electromagnetic fields, in flat spacetime [44, 45] and in generic curved $(1+1)$-dimensional spacetimes [46]. In fact, in $(3+1)$-dimensional spacetime, both non-trivial topology [44] and curvature [46] can induce violations of the averaged null energy condition. Moreover, in Wald and Yourtsever paper [46] it was shown that there are generic classes of spacetimes where quantum effects may violate the averaged null energy condition. Under these conditions, it is impossible today to exclude the possibility of such a violation of the averaged null energy condition as required for time machine creation. Also Flanagan and Wald [36] investigated whether the averaged null energy condition holds, for self-consistent solutions of the semi-classical Einstein equations

$$R_{\alpha\beta} - \frac{1}{2} g_{\alpha\beta} R = 8\pi \langle T_{\alpha\beta} \rangle, \tag{19}$$

where $\langle T_{\alpha\beta} \rangle$ is the expectation value of the renormalized stress-energy tensor of quantum fields. They investigated a free, linear, massless scalar field, with arbitrary curvature coupling, in the context of perturbation theory around the vacuum solution. They showed that probably, if traversable wormholes do exist as self-consistent solutions of the semi-classical equations, they cannot be macroscopic, but must be at Planck scale [see also Ford and Roman [47]]. So the question is still open for the case of general conditions. [See also [35]]. On the question of a stable Wormhole, a recent (2013) very interesting claim is due to Bronnikov, Lipatova, Novikov and Shatskiy [64], they presents a static, spherically symmetric wormhole model whose metric coincides with that of the so called Ellis wormhole, but the material source of gravity consists of a perfect fluid with negative density and a source-free radial electric or magnetic field.

For a certain class of fluid equations of state, this wormhole model is linearly stable under both spherically symmetric perturbations and axial perturbations of arbitrary multipolarity. A similar behavior is predicted for polar non spherical perturbations. It thus seems to be the first example of a stable wormhole model in the framework of general relativity (at least without invoking phantom thin shells as wormhole sources). In this paper they shown that, in General Relativity, it is really possible to construct a static traversable Lorentzian wormhole which is stable under all spherical and axial perturbations and it turns out to be independent of the equation of state assumed for the matter supporting the wormhole. The researches on stability, for different static, spherically symmetric configurations shows that non spherical polar perturbations behave qualitatively in the same way as their axial counterparts because, for both kinds of perturbations, the wave equations with nonzero multipolarities L possess effective potentials with positive "centrifugal" terms [1, 19]. Therefore if a model is really unstable, the instability will most probably manifest itself in spherical modes where $L = 0$. Nevertheless, to complete the stability study of that model, non spherical polar perturbations should be studied in deep. At the present that model is the best example of a stable wormhole model in general relativity. If there not exists stable traversable wormholes in the Universe, then in order to create a time machine by the present mechanism, one needs first to create such a wormhole. In the absence of wormholes, the space initially was simply connected and if a wormhole is created, then a later spacelike surface S' has different spatial topology. If the mechanism of formation of a time machine are restricted to a spatially bounded region, then it is possible to assume that one can surround this region by a timelike cylinder T , which intersects the spacelike surfaces S and S' in compact regions ST and $S'T$ of different topology. What happens is that the topology change occurs in a spacetime region $\mathcal{M}T$ bounded by S, S', and T. In absence of singularities and of the impossibility for $\mathcal{M}T$ to extend to infinity, $\mathcal{M}T$ is compact. Hawking [4] claimed that the change of spatial topology inside $\mathcal{M}T$ is impossible until it contains CTCs. This is a generalization of a well-known theorem by Geroch [48]. This result means that even the creation of a wormhole cannot be possible without the creation of a time machine, and it remains unclear, in the framework of pure General Relativity, if it is forbidden by physical laws,. But let discuss whether it is possible to have any reasonable physics in a spacetime with CTCs.

3.4. Time in Classical and Quantum Physics

The real problem of spacetimes with CTCs is the problem of causality. The existence of CTCs allows one to travel into the past and this at the end leads to the possibility of changing that past, so inevitably producing causality violations. In order to avoid contradictions in this approach, some authors estimate necessary to impose a principle of self-consistency. This principle was proposed and discussed in several works by Novikov [49, 51], Zeldovich and Novikov [50], Novikov and Frolov [37], Friedman *et al.*, [38]. The meaning of this is the following. For an open timelike curve, any event x divides the other events on this curve into two sets: future events and past events with respect to x. All past events can influence x, but future events not. On a CTC the choice of the event x divides other events on the curve into future events and past ones only locally. In this case, events, which locally are in the future with respect to x, can influence nevertheless the event x, going circularly around the CTC.

We must stress that there is no global division of events into future and past, on the CTCs. The future influences the present around the CTC, with a locally future-directed half of a light cone at each event of the closed curve. Not only the future is a result of evolution of the past, but the past is a result of the future also. All events in a spacetime with CTCs must be self-consistent. Following the principle of self-consistency, all events on CTCs influence each other around the closed timelike line in a self-adjusted way. A precise formulation of this principle is: *the only solution of the law of physics that can occur locally in the real Universe ore those which are globally self-consistent.* The principle of self-consistency forbids changing the past. All events happen only once, and especially cannot be changed, in such a way introducing a strong limitation on the meaning of time travels in the framework of standard General Relativity. In order to show how this principle works, several authors considered, in the proximity of the mouths of a wormhole, the so-called billiard ball problem (Friedman *et al.*, [38], Echeverria, Klinkhammer, and Thorne [52], Mikheeva and Novikov [53] which is the following : if a solid perfect1y elastic ball moves relative to the mouths of the wormhole, its speed is assumed to be small compared with the speed of light, so it can be treated non relativistically. But what happens is that the ball enters the wormhole through mouth 1, emerges from 2 in the past and, continuing its motion, it can encounter and collide with itself. At first look, there is a paradox in this problem. The starting position and velocity of the ball are chosen, so that the ball moves along a trajectory, enters mouth 1, and exits from mouth 2 earlier in time than it entered into 1. Then the ball continues its motion along a second trajectory. The timing is just right for the ball to hit itself at the point x, knocking its "younger" self along a third trajectory and thereby preventing itself from ever reaching mouth 1. This evolution is not self-consistent, and hence, in this approach, it is impossible. The mistake origin of the paradox is the following: when at the beginning of our scheme we continued the first trajectory after point x, we did not consider the influence of the impact and considered the motion of this ball after the passage of the mouths 1 and 2 along the second trajectory without taking into account this impact. Let's stress what this really means : this means that we did not take into account the influence of the future on the past. Now the emerging questions are: 1) Does a solution to this system exist for any initial conditions? 2) Is this solution unique? 3) What does the multiplicity of self-consistent solutions mean? 4) Does it have any physical meaning or does it simply indicate that the standard laws of physics cannot deal with time machines in a reasonable way?

In this context there is no definite answer. But we know that physics is quantum mechanical (not classical) by nature. If one considers the classical problem as a limit case of the quantum one, then the Cauchy problem turns out to be well posed in the formalism of Quantum Mechanics. Up to now we discussed plane motions of billiard balls. In classical physics more complicated interacting systems than the billiard ball have been investigated (for example, see Novikov [51], Lossev and Novikov [54]). Originally it was believed that, self-consistent evolution does not exist, but no clear demonstration of that have been showed. Carlini *et al.*, [55, 56], demonstrated that, for a billiard-ball problem, the principle of self-consistency is a consequence of the principle of least action, where the initial and final positions of balls are fixed. Going on and looking at Morris *et al.*, [8] and Friedman *et al.*, [38] it is possible to show that the chronology horizon is stable against classical perturbations. For quantum fields, due to the existence of zero-point fluctuations which cannot be suppressed, the renormalized vacuum stress-energy tensor $\langle T_{\mu\nu} \rangle$ diverges near the future chronology horizon, as shown by many authors [4, 35, 57] and quantum effects (vacuum polarization) directly are involved in the mechanism of time machine origin [58]. But there is another even more general problem. If a time machine exists, is it possible to formulate quantum mechanics in such a world? Although this problem is extremely complicated, the studies on that has proved to be productive. Foremost, in presence of a time machine, in General Relativity there is no well defined global Time. It comes out that it is impossible to determine a quantum state at given moment of time and to write the Schröedinger equation describing evolution of a quantum state. Nevertheless, the "Feynman sum over histories" can be used to obtain the amplitude of transitions (from one spacetime point to another). It happens that for non relativistic particles [59] and also for relativistic fields, this approach enables to compute unique probabilities for the outcomes of all measurements, that one might reasonably make, also in the achronal regions of spacetime. However, if the particles or fields are self-interacting, quantum theory in the presence of time machine is very peculiar. If a time machine was "operating" only during finite period, and there exist initial and final regions with the usual causal properties, the transition matrix from the initial region to the final region is non-unitary, for an interacting field theory calculated order by order in perturbation theory [59, 11]. One can have classical Hamiltonian quantum mechanics in the chronal future domain, but only under the condition that the future influences probabilities in the chronal initial region [60]. In this situation the approach based on the generalized quantum theories developed by Hartle [61] is appropriate. (For quantum theory in the presence of a time machine, see also Deutsch [62], Thorne [34], Goldwirth *et al.*, [63]). Frequently, it is implicitly assumed that, before the formation of chronology horizon, the standard local causal quantum field theory is valid. The situation would be completely different, if it happens that a non-contradictory quantum mechanics in a spacetime with CTCs, is possible only if there exist some global acausal prescriptions on the choice of states and their evolution, changing the standard equations not only in regions with CTCs, but also in the regions lying to the past of the chronology horizon. In this situation, the conclusions concerning the renormalized stress-energy tensor near the chronology horizon can obviously change. At the end a Time machine is maybe not excluded in classical General Relativity [64], even if it seems restricted with respect to the meaning which it evokes, but in any case it is in a theory able to take in account also Quantum Mechanics that significant progress can be expected.

4. Quantum Mechanics and General Relativity: The Open Quantum Relativity

The XXth century, the century of the greatest scientific realization, started apparently close to a general unifying vision of the physics, but ended still far from this goal. The successes of a sophisticated experimental physics, investigating also very far from our human environment (from remote galaxies to primordial particles), have brought surprising results very difficult to be settled in our standard schemes and, together with the failure in describing general relativity and quantum mechanics in the same theory, are at the basis of the lack of this unified vision. To achieve this goal, we introduce a formulation of gravitation in terms of symplectic structure [65, 66] (recovering a common feature with other interactions), and this, together with a General Conservation Principle taken into account in the symplectic formalism, seems able to solve shortcomings of modern physics. In fact despite the rapid progress, several new problems were substantially removed, but they are no longer forgettable if we want to carry on in this progress. Let us remember the principals of them: a) the contradiction pointed out by the Einstein-Podolsky-Rosen (EPR) paradox [67], i.e., the possibility, under particular conditions, to perturb a physical object without interacting with it in any known way; b) the existence of objects, like the black holes, which, despite serious attempts [68, 69], seem to violate the total energy conservation [70];
c) the consideration of quarks as elementary constituents even though it is generally hypothesized that probably they can never get their individuality [71]; d) the observed absence of primary antimatter in our universe, despite the standard symmetric creation of matter and antimatter pairs [72, 75]; e) the Big Bang theory, not satisfactory solved in the initial singularity [76]; f) the experimental results of quantum teleportation [77, 78] which suggest a $\Delta t = 0$ in transferring information, so questioning both the relativity and the causality; g) the lack of a consistent unitary description of all fundamental interactions [71]; h) the failure of the existent theories in describing cosmological evolution and cosmological parameters without introducing arbitrary and hidden quantities; i) the absence of mass generation theory fully resolved in it's dynamics; and, above all (*l*),

the absence of a general scheme [79, 80] connecting Quantum Mechanics and Relativity.
Now, according to us, several contradictions of contemporary physics arise from two principles, implicitly contained in every scientific formulation, principles certainly reasonable, but that are common sense assumptions and not mathematical theorems: i) the assumption of the existence of only one time arrow, from the past toward the future, which leads in several cases to a possible violation of the causality principle [79, 80]; ii) the assumption that the geometry is always given in the evolution of physical systems (as an example the invariance for diffeomorphisms) and any possible topology change is considered as a "singularity".
In other words, the topology of manifolds where systems evolve, is given from the beginning and never considered as a dynamical structure. In our opinion, these two principles are leading to singularities, symmetry breakings and violations of conservation laws, and are responsible for the most significant contradictions of today physics, since it seems that symmetry breakings and violations are a sort of ad hoc hypotheses invoked as soon as a new phenomenon cannot be included in standard schemes.
A way to take into account the above issues is deeply related to conservation laws, so then we propose to investigate what happens if conservation laws are always valid and symmetries are always maintained. As we will see, directly due to the fact that conservation laws can never be violated, the symmetry of the theory leads to the general consequence that backward and forward time evolution (and causation) are both allowed and, moreover, it emerges the necessity of a generalization to five physical dimensions which leads to the derivation of particle features as mass, spin, and charge as result of an embedding process. The foundations of quantum mechanics, in relations to relativity, are always largely discussed, since the intrinsic characteristic of quantum mechanics is the existence of systems that have not definite values of measurable quantities, unless one does not measure them. The most significant point is that this is due to fact that the state of a system is a superposition of different states, and the only possibility, left to the observation, consists in interacting in an irreversible way with the system, by the measure. This is also described as the collapse of the wave function by the measure, and then the evolution of the system, after the measure, has the collapsed state as initial wave function. This description, known as Copenhagen Interpretation (CI), is mainly due to Bohr, and, following it, the irreversibility of the measure process, and consequentially the irreversibility of

the time direction, plays a key role. Moreover the superposition of quantum states, that in CI is characteristic of the microscopic description of Nature, may have macroscopic effects and this is at the base of the EPR paradox and is the starting point of our approach. Basically, the Open Quantum Relativity is a theory based on two fundamental features: the assumption of a General Conservation Principle which states that the conservation laws can never be violated and the achievement that both General Relativity and Quantum Mechanics can be described under the standard of a covariant symplectic formalism. These facts lead to some general consequences. First of all the existence of a dynamical unification scheme of fundamental interactions [93] obtained by assuming a 5D space which allows that the conservation laws are always and absolutely valid as a natural necessity. Second the achievement that what we usually define as a singularity violating the conservation laws can be described instead by a process of topology change. Last the existence of a process of embedding and dimensional reduction, which gives rise to an induced gravity-matter theory in the 4D space-time, by which the usual masses, spins and charges of particles, naturally spring out. As final results the theory leads to a dynamical explanation of several problems of modern physics (e.g. entanglement of quantum states, quantum teleportation, gamma ray bursts origin, black hole singularities, cosmic primary antimatter absence). Moreover the theory provides a self-consistent picture of the observed accelerated cosmological behavior with the correct reproduction of experimental cosmological parameters. New predictions are also coming out from OQR, as gravitationally induced neutrino oscillations [176], indication of masses derived by this extended theory of Gravity [177], further scalar modes in gravitational waves [179] and determination of the scale of selfgravitating system by the upper limit of graviton mass [180]. The emergence of an extra force term in the reduction process and the possibility to recover the masses of particles, allow to reinterpret the Equivalence Principle as a dynamical consequence which naturally selects geodesics from metric structure and vice-versa the metric structure from the geodesics. It is important to stress that, in the Einstein General Relativity, geodesic structure is "imposed" by choosing a Levi-Civita connection and this fact can be criticized considering a more general completely "affine" approach like in the Palatini formalism (in agreement with our covariant symplectic approach). The dimensional reduction process gives rise to the generation of the masses of particles, which emerge both from the field equations and the embedded geodesics and, thank to that, the coincidence of chronological and geodesic structures is derived from the embedding and a new dynamical formulation of the Equivalence Principle is the direct consequence of dimensional reduction. The dynamically derived structure becomes more general, since two time arrows and closed time-like paths naturally emerge, so leading to fundamental consequences like a reinterpretation of the standard notion of causality, which can be now recovered even in the case in which it is questioned (like in entanglement phenomena and quantum teleportation), since it is generalized to a "forward" and a "backward" causation. At the end the Open Quantum Relativity is the formulation of a coherent dynamical theory, based on a principle: the General Conservation Principle, on a method: the dynamical deduction of the laws in an unitary frame with the minimal number of free variables and, finally, on a fundamental consideration: the possibility of a symplectic description for both Quantum Mechanics and General Relativity.

4.1. The General Conservation Principle: 5D-space Extension and 4D-Space Reduction

We begin now the description of the Open Quantum Relativity, theory able to solve several contradictions in the relations between Quantum Mechanics and Relativity [88, 89, 90], and we start with the introduction of the General Conservation Principle: "*The conservation laws have an absolute meaning and maintain always their validity. This is the reason why, when it would be otherwise impossible to preserve this validity, they determine the entanglement phenomenon, which allows in any case their recovery, thanks to topology changes with related time arrow inversions. So, the conservation laws are invariant while the topologies are not*" [85, 90]. This principle, thanks to the fact that mathematical formalism does not prevent backward and forward evolution in time, i) allows an explanation in terms of non-local physical behavior for entangled systems, ii) agrees with the experimentally observed violation of Bell inequalities and consequently of Einstein locality principle [79], iii) open a discussion on the first principles leading to such a symmetric time behavior.
For the "traditional" causality principle [79] and Einstein, Podolsky and Rosen point of view[67], the Quantum Mechanics lead to a paradox. Nevertheless, using backward and forward evolution, entangled systems are naturally explained [86].

In other words, if we maintain conservation laws absolutely valid, we need two time arrows, and backward evolution turns out to be a feature of Nature, which we cannot ordinarily feel, but which emerges as soon as a conservation law has no other way to maintain its validity. Strictly related to the concept of entanglement, there are topology changes taking place in order to preserve a conservation law, like matter-energy in black hole dynamics, or quantum numbers in EPR effects [86]. In the first case, as we will see in the following, the entangled system is constituted by a black hole which dynamically evolves in a white hole through a topology change (the essence of a worm hole).

The general feature of such a result is that the topologies are not invariant and so they can be considered dynamical quantities. The existence of backward and forward evolution leads however to the necessity of a fifth dimension. Let us sketch a very simple picture, imagining a hypothetical one-dimensional universe. In such a universe, a point placed to the left of another one could never invert its position exchanging the left with the right, while if we pass to a 2D-universe, it becomes obviously possible to exchange the relative position of the two. This is conceptually true also passing from four to five dimensions, allowing the evolution in both directions of the time axis [88, 89, 90]. But this is not just a technicality, because the 5D universe has a real physical meaning, in fact as the fourth dimension is related to time in Relativity, the fifth dimension can be related to the mass of particles, as we will see [90]. In other words, we can deal with a generalized 5D mass-chronotope, every point of which is labelled by space, time and mass. It is relevant to note the fact that we do not perceive the fourth time-like dimension in the same way of space-like dimensions and the situation is analogue for this fifth mass-like dimension. In previous papers, we discussed the approach for different cases. In [86], we showed that EPR paradox can be interpreted under the standard of conservation laws since the paradox is solved by an entangled superposition of forward and backward solutions. The set of equations (see Chap.5) coming out from the EPR effect (135) and (136) are fundamental in our approach, since they give rise to forward and backward dynamics. From a relativistic point of view, in the cases (135) and (136) (massive particles) we are inside one of the folds of the light-cone, in the case (137) (massless particles) we are, instead, on the null surface.

A role in distinguishing dynamics is played by the mass of particles which was introduced by the self-interacting potential (130). In a 5D-manifold, all particles are moving without preference of backward and forward evolution, while it is a reduction procedure from 5D to 4D-spacetime which gives rise to two dynamics (i.e., two arrows of time). In the light of these considerations, the concepts of entanglement and topology change are features of the theory which emerge in order to preserve conservation laws [85]. In this approach, the possibility to evolve backward to the past (i.e., in what we conceive as a past) is a general feature of Nature, but in our "forward" fold of light-cone, we cannot sensorially perceive it. The only way to experience the backward evolution is under the extreme conditions in which a conservation law, being otherwise violated, determines a spacetime breaking via a topology change connecting entangled systems and this phenomenon is allowed by the presence of a fifth dimension of real physical meaning. We want to outline that this is a general result, because the connection among conservation laws, symmetries and first integrals of motion have a deep physical meaning. In view of that, a 5D-theory can be the minimal dimensional unification scheme which, by a reduction to 4D,can give rise to the physics which we experimentally know (see also [91]). Beside a dynamical unification scheme of all the interactions [93], we develop an approach by which, starting from a pure theory of gravitation formulated in 5D, we are capable of obtaining 4D-physics where the ordinary features of particles, as mass, spin and charge, are the product of an embedding process, as well as the same structure of spacetime, which results richer than the standard 4D-one, since backward and closed timelike solutions naturally emerges. An important feature of the approach in which conservation laws are always valid, is that the 5D-Bianchi identities always hold [90], and what we consider as different symmetries and interactions can be all reconducted to a 5D-supergroup, whose possible 4D decomposition, including the Standard Model, [93] is

$$\mathcal{G}_5 \supset IO(3,1) \otimes SU(3) \otimes SU(2) \otimes U(1). \tag{20}$$

So we are going to deal with the degrees of freedom of the spacetime and of the particles under the same standard, and $\mathcal{G}_5 = SL(5)$ is the minimal group which is capable of including all the fundamental interactions and the 10 generators of inhomogeneous Lorentz group [93].

4.2. From 5D to 4D Field Equations. Geodesics, Embedding and Mass Generation

Here we want to present the main features of this 5D-theory discussing, in particular, the 4D reduction procedure which induces a scalar-tensor theory of gravity where conservation laws (i.e., Bianchi identities) play a fundamental role into dynamics. After that we develop the related quantum cosmology showing that, using these results, it is always possible to select classical trajectories, i.e., observable universes. But let see. The 5D-manifold which we are considering is a Riemannian space provided with a 5D-metric of the form

$$dS^2 = g_{AB} dx^A dx^B \tag{21}$$

where the Latin indexes are A, B = 0, 1, 2, 3, 4. We do not specify the signature yet, since it can be dynamically fixed by the reduction procedure in 4D. Canonically, it is a pseudo-Lorentzian metric. We define the curvature invariants, the field equations and the conservation laws in the 5D-space. Asking for a space which is a smooth manifold and singularity free, because defined in such a way that every conservation law on it has to be always and absolutely valid [93]. The 5D-Riemann tensor is

$$R^D_{ABC} = \partial_B \Gamma^D_{AC} - \partial_C \Gamma^D_{AB} + \Gamma^D_{EB} \Gamma^E_{AC} - \Gamma^D_{EC} \Gamma^E_{AB}. \tag{22}$$

The number of independent components of such a tensor, after the full derivation and thanks to the Petrov classification [92], is

$$\frac{1}{12} N^2 (N^2 - 1) = 50 \quad for \quad N = 5 \tag{23}$$

The Ricci tensor and scalar are derived from the contractions

$$R_{AB} = R^C_{ACB}, \quad {}^{(5)}R = R^A_A. \tag{24}$$

The field equations can be derived from the 5D Hilbert-Einstein action

$${}^{(5)}A = -\frac{1}{16\pi {}^{(5)}G} \int d^5x \sqrt{-g^{(5)}} \, [{}^{(5)}R], \tag{25}$$

where ${}^{(5)}G$ is the 5D-gravitational coupling and $g^{(5)}$ is the determinant of the 5D-metric. The variational principles

$$\delta \int d^5x \sqrt{-g^{(5)}} \, [{}^{(5)}R] = 0, \tag{26}$$

gives te 5-D field equations which are

$$G_{AB} = R_{AB} - \frac{1}{2} \, g_{AB} \, {}^{(5)}R = 0, \tag{27}$$

so that at least the Ricci-flat space is always a solution. Let us define now the 5D stress-energy tensor:

$$T_{AB} = \nabla_A \Phi \nabla_B \Phi - \frac{1}{2} g_{AB} \nabla_C \Phi \nabla^C \Phi \tag{28}$$

where, because of the General Conservation Principle and so the presence of symmetries, only the kinetic terms are considered. As standard, such a tensor can be derived from a variational principle

$$T^{AB} = \frac{2}{\sqrt{-g^{(5)}}} \frac{\delta(\sqrt{-g^{(5)}} \mathcal{L}_\Phi)}{\delta g_{AB}} \tag{29}$$

where $\mathcal{L}_\Phi$ is a Lagrangian density connected with the scalar field Φ. Because of the definition of 5D space itself [90] it is important to stress now that no self-interaction potential $V(\Phi)$ has been taken into account so that T_{AB} is a completely symmetric object and Φ is, by definition, a cyclic variable. This fact guarantees that Noether theorem always holds for T_{AB} and so a conservation law intrinsically exists. The field equations can now assume the form

$$R_{AB} = \chi(T_{AB} - \frac{1}{2} g_{AB} T), \tag{30}$$

where T is the trace of T_{AB} and $\chi = 8\pi^{(5)}$G, being $\hbar = c = 1$. The form (30) of field equations is useful to put in evidence the role of the scalar field Φ, if we are not simply assuming Ricci flat 5D-spaces. As we said, T_{AB} is a symmetric tensor for which holds the relation

$$T_{[AB]} = T_{AB} - T_{BA} = 0 \tag{31}$$

Due to the choice of the metric and to the symmetric nature of the stress-energy tensor T_{AB} and the Einstein field equations G_{AB}, the contracted Bianchi identities

$$\nabla_A T^A_B = 0, \quad \nabla_A G^A_B = 0 \tag{32}$$

must always hold.
Developing the stress-energy tensor, we have

$$\nabla_A T^A_B = \nabla_A(\partial_B \Phi \partial^A \Phi - \frac{1}{2} \delta^A_B \, \partial_C \, \Phi \partial^C \Phi)$$

$$= (\nabla_A \Phi_B)\Phi^A + \Phi_B(\nabla_A \Phi^A) - \frac{1}{2}(\nabla_B \Phi_C)\Phi^C - \frac{1}{2}\Phi_C(\nabla_B \Phi^C)$$

$$= (\nabla_A \Phi_B)\Phi^A + \Phi_B(\nabla_A \Phi^A) - \Phi_C(\nabla_B \Phi^C) \tag{33}$$

Since our 5D-space is a Riemannian manifold, it is

$$\nabla_A \Phi_B = \nabla_B \Phi_A \tag{34}$$

and then

$$\Phi^A(\nabla_A \Phi_B) - \Phi_C(\nabla_B \Phi^C) = \Phi^A(\nabla_B \Phi_A) - \Phi_C(\nabla_B \Phi^C) = 0. \tag{35}$$

In this case, partial and covariant derivatives coincide for the scalar field Φ. So then

$$\nabla_A T^A_B = \Phi_B \,{}^{(5)}\Box\Phi \tag{36}$$

where ${}^{(5)}\Box$ is the 5D d'Alembert operator defined as

$$\nabla_A \Phi^A \equiv g^{AB}\Phi_{,A;B} \equiv {}^{(5)}\Box\Phi \tag{37}$$

At the end the conservation of the stress-energy tensor T_{AB} (i.e., the contracted Bianchi identities) implies the Klein-Gordon equation which assigns the dynamics of Φ, that is

$$\nabla_A T^A_B = 0 \quad \Longleftrightarrow \quad {}^{(5)}\Box\Phi = 0 \tag{38}$$

assuming $\Phi_B \neq 0$ since we are dealing with a real physical field. This field equations, are providing a physical meaning to the fifth dimension. Splitting the 5D problem in a $(4+1)$ problem, from this Eq. (38), it is possible to generate the mass of particles in 4D [89, 90, 93]. Let us now derive and discuss the geodesic equation in this 5D-manifold. The action for geodesics is

$$\mathcal{A} = \int dS \Big(g_{AB} \frac{dx^A}{dS}\frac{dx^B}{dS} \Big)^{1/2} \tag{39}$$

the Euler-Lagrange equations give the geodesic equation

$$\frac{d^2 x^A}{dS^2} + \Gamma^A_{BC}\frac{dx^B}{dS}\frac{dx^C}{dS} = 0 \tag{40}$$

where Γ^A_{BC} are the 5D-Christoffel symbols. Eq. (40) can be split in the $(4+1)$ form

$$2g_{\alpha\mu}\Big(\frac{dx^\alpha}{ds}\Big)\Big(\frac{d^2x^\mu}{ds^2} + \Gamma^\mu_{\beta\gamma}\frac{dx^\beta}{ds}\frac{dx^\gamma}{ds}\Big) + \frac{\partial g_{\alpha\beta}}{ds^4}\frac{dx^4}{ds}\frac{dx^\alpha}{ds}\frac{dx^\beta}{ds} = 0 \tag{41}$$

where the Greek indexes are $\mu, \nu = 0, 1, 2, 3$ and $ds^2 = g_{\alpha\beta}dx^\alpha dx^\beta$

In the 4D reduction (i.e., in the usual spacetime) we see only the standard geodesics of General Relativity, i.e., the 4D-component of Eq. (41)

$$\frac{d^2x^\mu}{ds^2} + \Gamma^\mu_{\beta\gamma}\frac{dx^\beta}{ds}\frac{dx^\gamma}{ds} = 0 \tag{42}$$

so that, the last part of the representation given by Eq.(41) has to vanish in 4D. For standard laws, the metric $g_{\alpha\beta}$ does not depend on x^4 in the embedded 4D-manifold, so the last component of Eq. (41) can be read as an "extra force" which gives the motion of a 4D-frame with respect to the fifth coordinate x^4. This fact shows that the fifth dimension has a real physical meaning and any embedding procedure scaling up in 5D-manifold (or reducing to 4D-spacetime) has a dynamical description [31]. The quantity

$$\mathcal{F} = \frac{\partial g_{\alpha\beta}}{dx^4}\frac{dx^4}{ds}\frac{dx^\alpha}{ds}\frac{dx^\beta}{ds} \tag{43}$$

has to be related to the mass of moving particles, being the extra-force related to the embedding procedure. The emergence of this term in Eq. (41), means that the Equivalence Principle in 4D can be seen as a dynamical result [91] and the quantity $\mathcal{F}$, which distinguishes the masses of particles, gives a dynamical base of the Equivalence Principle. Finally all particles are massless in 5D while, for the physical meaning of the fifth coordinate, they acquire mass in 4D thanks to Eq. (43). Another important consideration has to be done on the line element ds^2. Let us take into account a 5D-null path. It is given by

$$dS^2 = g_{AB}dx^A dx^B = 0 \tag{44}$$

A splitting of Eq. (44), made considering the 4D-part of the metric and the homogeneous 5D component, is

$$dS^2 = ds^2 + g_{44}(dx^4)^2 = 0 \tag{45}$$

consequently a null path in 5D can result a pure time-like or a space-like path in 4D depending on the sign of g_{44}. Let us take into account the 5D-generalized velocity $u^A = dx^A/ds$. It can be split as a velocity in the ordinary 3D-space v, a velocity along the ordinary time axis w and a velocity along the fifth dimension z. In general, for null paths, we can have $\nu^2 = w^2 + z^2$ and this should lead to super-luminal speed, explicitly overcoming the Lorentz transformations.

The problem is solved if we consider the 5D-motion as a-luminal, because all particles (massless) and fields have the same speed and the distinction among super-luminal motion, luminal motion and sub-luminal (or canonically causal motion for massive particles) emerges only after the dynamical reduction from 5D to 4D-spacetime. In this way, the fifth dimension, is generating the different dynamics which we conceive in 4D and this process of mass generation sets the particles in the 4D-light-cone. The Embedding process, where the choice of 5D coordinates plays the role of Gauge specifying the 4D physics (also in the non conventional entanglement), give rise to the masses and vice-versa the bijective embedding gives physical meaning to the fifth coordinate x^4. By replacing the variational principle (26) with

$$\delta \int d^{(5)}x\sqrt{-g^{(5)}}[^{(5)}R + \lambda(g_{44} - \epsilon\Phi^2)] = 0 \tag{46}$$

where λ is a Lagrange multiplier, Φ a generic scalar field and $\epsilon = \pm 1$.
From eq. (46) we can write

$$dS^2 = g_{AB}dx^A dx^B = g_{\alpha\beta}dx^\alpha dx^\beta + g_{44}(dx^4)^2 = g_{\alpha\beta}dx^\alpha dx^\beta + \epsilon\Phi^2(dx^4)^2 \tag{47}$$

Let us now consider a flat 5D-space. The d'Alembert operator

$$\Box = \partial^\mu \partial_\mu \tag{48}$$

in 5D can be split, following the metric definition (47) for particle-like solutions, as:

$$^{(5)}\Box = \Box - {\partial_4}^2 \tag{49}$$

so selecting the value $\epsilon = -1$ in the metric. Now, introducing the scalar field Φ, we have

$$^{(5)}\Box\Phi = [\Box - {\partial_4}^2]\Phi = 0 \tag{50}$$

and then

$$\Box\Phi = {\partial_4}^2\Phi \tag{51}$$

The problem is solvable by separation of variables and then we split the scalar field Φ into two functions

$$\Phi = \phi(t, \vec{x})\chi(x_4) \tag{52}$$

where the field ϕ depends on the ordinary space-time coordinates, while χ is a function of the fifth coordinate x_4. Inserting (52) into Eq.(51), we get

$$\frac{\Box\phi}{\phi} = \frac{1}{\chi}\left[\frac{d^2\chi}{dx_4^2}\right] = -k_n^2 \tag{53}$$

where k_n must be a constant for consistency. From Eq.(53), we obtain the two equations of motion

$$(\Box + k_n^2)\phi = 0 \tag{54}$$

and

$$\frac{d^2\chi}{dx_4^2} + k_n^2\chi = 0 \tag{55}$$

Eq.(55) describes a harmonic oscillator whose general solution is

$$\chi(x_4) = c_1 e^{-ik_n x_4} + c_2 e^{ik_n x_4} \tag{56}$$

The constant k_n has the physical dimension of the inverse of a length and, assigning the boundary conditions for consistency with (55), we can derive the eigenvalue relation

$$k_n = \frac{2\pi}{l}\, n, \tag{57}$$

where n is an integer and l a length which we have previously defined in Eq. (55). As a result, in standard units, we can recover the physical lengths through the Compton lengths

$$\lambda_n = \frac{\hbar}{2\pi m_n c} = \frac{1}{k_n} \tag{58}$$

which always assign the mass of a particle. It has to be emphasized that, the eigenvalues of Eq.(55) are the masses of particles which are generated by the process of reduction from 5D to 4D, Eqs. (50), (51). The solution (56) is the superposition of two mass eigenstates and the 4D evolution is given by Eq. (54). An important point to be stressed is that, due to the fact that we have a harmonic oscillator Eq. (55), we have that the solutions of this one give us the associated Compton lengths from which the effective physical masses are derived. More precisely, different values of n fix the families of particles, while, for any given value n, different values of parameters $c_{1,2}$ distinguish the different particles within a family.

Furthermore, the effective mass can be geometrically derived as

$$m_{eff} \equiv \int |\Phi| dx^4 = \int |\Phi(dx^4/ds)| ds \tag{59}$$

where ds is the affine parameter used in the above derivation of geodesic equation. The above reduction procedure from 5D to 4D tells us that, due to the coincidence of the descriptions (56) and (59), this scheme is an effective mechanism for mass generation.

4.3. The Symplectic Structure of Open Quantum Relativity

As we have seen, it seems necessary the existence of a link between General Relativity and Quantum Mechanics and we will show that it exists an intrinsic common symplectic structure in General Relativity and Q.M. which can be recovered from conservation laws in Hamiltonian formulation. A self-consistent quantum field theory of space-time has not been achieved, up to date, using standard quantization approaches. In fact, the request of general coordinate invariance of General Relativity gives rise to several shortcomings in quantum field theory. For a generic physical field, one has to assign the field amplitudes and their first time derivatives, in order to determine the time evolution of such a field considered as a dynamical entity, but in General Relativity, these quantities are not suitable for dynamical determination, since the metric $g_{\alpha\beta}$ can evolve at any time by a coordinate transformation. This means that it is necessary a separation of metric degrees of freedom into a part related to the true dynamical information and a part related to the coordinate system. This operational approach works in Electromagnetism while it fails in General Relativity, where dynamics is highly non-linear. Independent dynamical modes of gravitational field are recovered when theory is formulated in a canonical form and the canonical formalism is preliminary in quantization program, since it gives rise directly to Poisson brackets. To realize such a program, one needs first order field equations and a $(3+1)-$form of dynamics. In General Relativity, the ADM formalism [141] has led to the definition of gravitational Hamiltonian and time as a conjugate couple of variables. In any case, the main feature of General Relativity, the covariance, is impaired and the full quantization of gravity was not achieved up to now. The main problems was related to the lack of Hilbert space and of a quantum measure for the metric. A prominent role [65, 66], in the identification of a covariant symplectic structure, can be played by bilinear Hamiltonians which have to be conserved. In fact, taking into account generic Hamiltonian invariants, constructed by covariant vectors, bivectors or tensors, it is possible to show that a symplectic structure can be achieved in any case. By specifying the nature of such vector fields (or, in general, tensor invariants), it gives rise to intrinsically symplectic structure which is always related to Hamilton-like equations (and a Hamilton-Jacobi-like approach is always found). This works for curvature invariants, Maxwell and Dirac fields. In any case, the two main basic assumptions are that conservation laws always have to be identified and moreover that the structure is completely affine and not metric. Such a program of "generalized" quantization works for every generic fields, From a theoretical point of view, such a covariant symplectic structure constitutes the core of Open Quantum Relativity. Here we outline that main features of a generic symplectic structure and relate it to the Hamiltonian mechanics, with the goal to achieve a covariant formulation. In general, a symplectic structure [65] is achieved if the couple

$$\{\mathbf{E_{2n}}, \mathbf{w}\}, \tag{60}$$

is defined, where E_{2n} is a vector space and the tensor w, on E_{2n}, associates scalar functions to pairs of vectors, that is

$$[\mathbf{x}, \mathbf{y}] = \mathbf{w}(\mathbf{x}, \mathbf{y}), \tag{61}$$

which is the antiscalar product. Such an operation satisfies the following properties

$$[\mathbf{x}, \mathbf{y}] = -[\mathbf{y}, \mathbf{x}] \qquad \forall \mathbf{x}, \mathbf{y} \in \mathbf{E_{2n}} \tag{62}$$

$$[\mathbf{x}, \mathbf{y} + \mathbf{z}] = [\mathbf{x}, \mathbf{y}] + [\mathbf{x}\mathbf{z}] \qquad \forall \mathbf{x}, \mathbf{y}, \mathbf{z} \in \mathbf{E_{2n}} \tag{63}$$

$$\mathbf{a}[\mathbf{x}, \mathbf{y}] = [\mathbf{a}\mathbf{x}, \mathbf{y}] \qquad \forall \mathbf{a} \in R, \quad \mathbf{x}, \mathbf{y} \epsilon \mathbf{E_{2n}} \tag{64}$$

$$[\mathbf{x}, \mathbf{y}] = 0 \qquad \forall \mathbf{y} \in \mathbf{E_{2n}} \Rightarrow \mathbf{x} = 0 \tag{65}$$

$$[\mathbf{x}, [\mathbf{y}, \mathbf{z}]] + [\mathbf{y}, [\mathbf{z}, \mathbf{x}]] + [\mathbf{z}, [\mathbf{x}, \mathbf{y}]] = 0 \tag{66}$$

The last one is the Jacobi cyclic identity. If $\{e_i\}$ is a vector basis, in E_{2n} the antiscalar product is completely singled out by the matrix elements

$$w_{ij} = [\mathbf{e_i}, \mathbf{e_j}], \tag{67}$$

where w is an antisymmetric matrix with determinant different from zero. Every antiscalar product between two vectors can be expressed as

$$[\mathbf{x}, \mathbf{y}] = w_{ij} x^i y^j \tag{68}$$

where x^i and y^j are the vector components in the given basis. The form of the matrix w and the relation (68) become considerably simpler if a canonical basis is taken into account for w. Since w is an antisymmetric non-degenerate tensor, it is always possible to represent it through the matrix

$$J = \begin{pmatrix} 0 & I \\ -I & 0 \end{pmatrix} \tag{69}$$

where I is a $(n \times n)$ unit matrix. Every basis where w can be represented through the form (69) is a symplectic basis. In other words, the symplectic bases are the canonical bases for any antisymmetric non-degenerate tensor w and can be characterized by the following conditions:

$$[\mathbf{e}_i, \mathbf{e}_j] = 0, [\mathbf{e}_{n+i}, \mathbf{e}_{n+j}] = 0, \qquad [\mathbf{e}_i, \mathbf{e}_{n+j}] = \delta_{ij}, \tag{70}$$

which have to be verified for every pair of values i and j ranging from 1 to n. Finally, the expression of the antiscalar product between two vectors, in a symplectic basis is

$$[\mathbf{x}, \mathbf{y}] = \sum_{i=1}^{n} (x^{n+1} y^i - x^i y^{n+1}) \tag{71}$$

and a symplectic transformation in E_{2n} leaves invariant the antiscalar product

$$\mathbf{S}[\mathbf{x}, \mathbf{y}] = [\mathbf{S}(\mathbf{x}), \mathbf{S}(\mathbf{x})] = [x, y]. \tag{72}$$

It is remarkable that standard Quantum Mechanics satisfies such properties and so it is endowed with a symplectic structure. Leaving to the quoted paper [75] for the full treatment, finally it comes out that

$$[U, V] = \frac{\partial U}{\partial x^\mu} \frac{\partial V}{\partial p_\mu} - \frac{\partial V}{\partial x^\mu} \frac{\partial U}{\partial p_\mu}, \tag{73}$$

where $U(q^\mu, \pi_\nu)$ and $V(q^\mu, \pi_\nu)$ are arbitrary function , which means that the fundamental Poisson brackets are always conserved. On the other hand, this implies that the variables $\{x^\mu, p_\nu\}$ are a canonical conjugate pair. As a general remark, we have to say that a generic metric $g_{\alpha\beta}$ and a connection $\Gamma^\alpha_{\mu\nu}$ are related to the fact that we are passing from a Minkowski-flat spacetime (local inertial reference frame) to an accelerated reference frame (curved spacetime). In what follows, we want to show that a generic bilinear Hamiltonian invariant, which is conformally conserved, is always based on a canonical symplectic structure. Summarizing, the main points which we need are: i) an even-dimensional vector space equipped with an antiscalar product satisfying the algebra (62)-(66); ii) generic vector fields defined on such a space which have to satisfy the Poisson brackets; iii) first-order equations of motion which can be read as Hamilton-like equations; iv) and, finally, the feature that general covariance has to be preserved.

Such a program can be pursued by taking into account covariant and contravariant vector fields. In fact, it is always possible to construct the Hamiltonian invariant

$$\mathcal{H} = V^{\alpha}V_{\alpha} \tag{74}$$

which satisfies the relation

$$\delta\mathcal{H} = \delta(V^{\alpha}V_{\alpha}) = 0, \tag{75}$$

being δ a spurious variation due to the transport. We must outline that the vectors V^{α} and V_{α} are not specified and all the following considerations are completely general. The Eq. (74) is an already parameterized invariant which can be seen as the density of a parameterized action principle [142, 143] where the time coordinate is not distinguished a priori from the other coordinates. The intrinsic variation of V_{α} is

$$DV^{\alpha} = dV^{\alpha} - \delta V^{\alpha} = \partial_{\beta}V^{\alpha}dx^{\beta} - \delta V^{\alpha} \tag{76}$$

where d is the total variation and δ the spurious variation. In General Relativity, if the spurious variation of a given quantity is equal to zero, it means that the quantity is conserved. Finally ([75] for the full development), we arrive to the formulation

$$\frac{\delta\mathcal{H}}{dx^{\beta}} - (\Gamma^{\alpha}_{\sigma\beta} - \Gamma^{\alpha}_{\sigma\beta})\left(\frac{\partial H}{\partial V_{\sigma}}\right)\left(\frac{\partial H}{\partial V^{\alpha}}\right) = 0 \tag{77}$$

identically equal to zero, so then $\mathcal{H}$ is absolutely conserved, and the analogy with a canonical Hamiltonian structure is straightforward. In fact, if

$$\mathcal{H} = \mathcal{H}(p, q) \tag{78}$$

is a classical generic Hamiltonian function, expressed in the canonical phase-space variables {p, q}, the total variation, in a vector space E_{2n} whose dimensions are generically given by p_i and q_j with $i, j = 1, ..., n$, is

$$d\mathcal{H} = \frac{\partial \mathcal{H}}{\partial q}\, dq + \frac{\partial \mathcal{H}}{\partial p}\, dp, \tag{79}$$

and

$$\begin{aligned} \frac{d\mathcal{H}}{dt} &= \frac{\partial \mathcal{H}}{\partial q}\dot{q} + \frac{\partial \mathcal{H}}{\partial p}\dot{p} \\ &= \frac{\partial \mathcal{H}}{\partial q}\frac{\partial \mathcal{H}}{\partial p} - \frac{\partial \mathcal{H}}{\partial p}\frac{\partial \mathcal{H}}{\partial q} \equiv 0, \end{aligned} \tag{80}$$

thanks to the Hamilton canonical equations

$$\dot{q} = \frac{\partial \mathcal{H}}{\partial p}, \qquad \dot{p} = -\frac{\partial \mathcal{H}}{\partial q}. \tag{81}$$

Such a canonical approach holds also in our covariant case if we operate the substitutions

$$V^{\alpha} \longleftrightarrow p \qquad V_{\alpha} \longleftrightarrow q \tag{82}$$

and the canonical equations are

$$\frac{\delta V^{\alpha}}{\partial x^{\beta}} = -\Gamma^{\alpha}_{\sigma\beta}\left(\frac{\partial \mathcal{H}}{\partial V_{\sigma}}\right) \longleftrightarrow \frac{dp}{dt} = -\frac{\partial \mathcal{H}}{\partial q}, \tag{83}$$

$$\frac{\delta V_{\alpha}}{dx^{\beta}} = \Gamma^{\alpha}_{\sigma\beta}\left(\frac{\partial \mathcal{H}}{\partial V^{\sigma}}\right) \longleftrightarrow \frac{dq}{dt} = \frac{\partial \mathcal{H}}{\partial p}. \tag{84}$$

In other words, starting from the (Hamiltonian) invariant (74), we have recovered a covariant canonical symplectic structure. The variation (75), once developed [75], may be seen as the generating function $\mathcal{G}$ of canonical transformations where the generators of q, p and t changes are dealt under the same standard. The covariant and contravariant vector fields can be also of different nature so that the above fundamental Hamiltonian invariant can be generalized as

$$\mathcal{H} = W^{\alpha}V_{\alpha}, \tag{85}$$

or, considering scalar smooth and regular functions, as

$$\mathcal{H} = f(W^{\alpha}V_{\alpha}). \tag{86}$$

or, in general

$$\mathcal{H} = f(W^{\alpha}V_{\alpha}, B^{\alpha\beta}C_{\alpha\beta}, B^{\alpha\beta}V_{\alpha}V'_{\beta}.....), \tag{87}$$

where the invariant can be constructed by covariant vectors, bivectors and tensors.

Clearly, the identifications

$$W^{\alpha} \longleftrightarrow p \qquad V_{\alpha} \longleftrightarrow q \tag{88}$$

hold and the canonical equations are

$$\frac{\delta W^{\alpha}}{dx^{\beta}} = -\Gamma^{\alpha}_{\sigma\beta}\Big(\frac{\partial\mathcal{H}}{\partial V_{\sigma}}\Big) \qquad \frac{\delta V_{\alpha}}{dx^{\beta}} = \Gamma^{\alpha}_{\sigma\beta}\Big(\frac{\partial\mathcal{H}}{\partial W^{\sigma}}\Big). \tag{89}$$

At the end, conservation laws are given by

$$\frac{\delta\mathcal{H}^{\alpha}}{dx^{\beta}} = (\Gamma^{\alpha}_{\sigma\beta} - \Gamma^{\alpha}_{\sigma\beta})\Big(\frac{\partial\mathcal{H}}{\partial V_{\sigma}}\Big)\Big(\frac{\partial\mathcal{H}}{\partial W^{\alpha}}\Big) \equiv 0 \tag{90}$$

This means that, in our scheme, the canonical symplectic structure is assigned in the way in which covariant and contravariant vector fields are related. If the Hamiltonian invariant is constructed by bivectors and tensors, Eqs. (89) and (90) have to be generalized but the structure is the same. We have to stress that, in our derivation, we never used the metric but only connections, considering a completely affine Palatini approach [107]. These considerations can be made independent of the reference frame, if we define a suitable system of unitary vectors by which we can pass from holonomic to anholonomic description and viceversa. We can define the reference frame, on the event manifold $\mathcal{M}$, as vector fields $e_{(k)}$ in event space and dual forms $e^{(k)}$ such that vector fields $e_{(k)}$ define an orthogonal frame at each point and the relations

$$e^{(k)}(e_{(l)}) = \delta^{(k)}_{(l)} \tag{91}$$

hold. If vectors are unitary, in a Riemann 4-spacetime, they are the standard vierbeins [24]. If we do not limit this definition of reference frame by orthogonality, we can introduce a *coordinate reference frame* $(\partial_{\alpha}, ds^{\alpha})$ based on vector fields tangent to line x^{α} = const.

Both reference frames are linked by the relations

$$e_{(k)} = e^{\alpha}_{(k)}\partial_{\alpha} \quad ; \qquad e^{(k)} = e^{(k)}_{\alpha}dx^{\alpha}. \tag{92}$$

Greek indices indicate holonomic coordinates while Latin indices between brackets, the anholonomic coordinates. We can prove the existence of a reference frame using the orthogonalization procedure at every point of spacetime and from the same procedure, we get that coordinates of frame smoothly depend on the point. The existence of a global reference frame follows from this. A smooth field on time-like vectors of each frame defines congruence of lines that are tangent to this field. We say that each line is a *world line of an observer or a local reference frame. Therefore a reference frame is a set of local reference frames.* The Lorentz transformation can be defined as a transformation of a reference frame

$$x'^{\alpha} = f(x^0, x^1, x^2, x^3, ..., x^n) \tag{93}$$

$$e'^{\alpha}_{(k)} = A^{\alpha}_{\beta} B^{(l)}_{(k)} e^{\beta}_{(l)} \tag{94}$$

where

$$A^{\alpha}_{\beta} = \frac{\partial x'^{\alpha}}{\partial x'^{\beta}}, \qquad \delta_{(i)(l)} B^{(i)}_{(j)} B^{(l)}_{(k)} = \delta_{(j)(k)}. \tag{95}$$

A vector field V has two types of coordinates: holonomic coordinates V^{α} relative to a coordinate reference frame and anholonomic coordinates $V^{(k)}$relative to a reference frame. We call the transformation A^{α}_{β} the holonomic part and transformation $B^{(l)}_{(k)}$ the anholomic. For these two kinds of coordinates, the relation

$$V^{(k)} = e^{(k)}_{\alpha} V^{\alpha} \tag{96}$$

holds. We can study parallel transport of vector fields using any form of coordinates. Because Eqs. (93) and (94) are linear transformations, we expect that parallel transport in anholonomic coordinates has the same form as in holonomic coordinates. Living to reference [75] for the complete calculation, all the above considerations can be immediately achieved in holonomic and anholonomic formalism considering the equation

$$\mathcal{H} = W^{\alpha} V_{\alpha} = W^{(k)} V_{(k)} \tag{97}$$

and the analogous ones. This means that the results are independent of the reference frame and the symplectic covariant structure always holds. In conclusion we have shown that a covariant, symplectic structure can be found for every bilinear Hamiltonian invariant, if these invariants are always conserved, as asked by the General Conservation Principle [90]. In particular, the Poisson brackets are always covariantly conserved for generic vector fields. Besides, the approach can be formulated in a holonomic and anholonomic representations, once vector, tensor and spinor fields are represented both in vierbein and coordinate-frames, feature essential to be sure that general covariance and symplectic structure are always conserved. Another important point is the role of affine connections, which are considered independently from metric. Due to this feature, the approach is genuinely Palatini-like since no Levi-Civita connection is requested. Furthermore, covariant and contravariant vector, tensor and spinor fields can be read as the configurations q^i and the momenta p_i of standard Hamiltonian dynamics in such a way that specific Hamilton field equations are recovered from the covariant derivatives of such covariant and contravariant vector fields. We have applied the method to gravity, electromagnetism and Dirac fields. For example, the curvature invariants of General Relativity can be recovered from Hamiltonian invariants opportunely defined, so that we must stress that this theory is essentially covariant and symplectic. If the vector fields are the 4-velocity in a generic curved space-time, we obtain the equation of geodesic motion and the geodesic deviation, which can be read as covariant Hamilton equations. If we consider the vector potential of Electromagnetism, Maxwell equations and Lorentz gauge are recovered and also in this case, these two sets of equations (Maxwell ones and Lorentz gauge) can be read as two sets of conjugate Hamilton equations. Finally, we discussed the mass-shell conditions of relativistic particle mechanics, which can be intrinsically considered as a bilinear Hamilton invariant. We will show that it gives rise to Klein-Gordon equation and Dirac equation when scalar or spinor fields are considered. The procedure can be generalized to any space-time and this is an essential feature of Open Quantum Relativity.

4.4. Symplectic–covariant Formalism: Properties and Applications

The symplectic approach is very general, as we will see from many applications.The invariant (74) is a generic conserved quantity specified by the choice of V^α and V_α. For example, if

$$V^\alpha = \frac{dx^\alpha}{ds}, \tag{98}$$

is a 4-velocity, with $\alpha = 0, 1, 2, 3$, from equation (83), we obtain the equation of geodesics

$$\frac{d^2x^\alpha}{ds^2} + \Gamma^\alpha_{\mu\nu}\frac{dx^\mu}{ds}\frac{dx^\nu}{ds} = 0 \tag{99}$$

On the other hand, being

$$\delta V^\alpha = R^\alpha_{\beta\mu\nu} V^\beta dx_1^\mu dx_2^\nu, \tag{100}$$

the result of the transport along a closed path, it is easy to recover the geodesic deviation considering the geodesic (99) and the infinitesimal variation with respect to it, i.e.,

$$\frac{d^2(x^\alpha + \xi^\alpha)}{ds^2} + \Gamma^\alpha_{\mu\nu} + (x+\xi)\ \frac{d(x^\mu + \xi^\mu)}{ds}\ \frac{d(x^\nu + \xi^\nu)}{ds} = 0 \tag{101}$$

which gives, through the Riemannian tensor,

$$\frac{d^2\xi^\alpha}{ds^2} = R^\alpha_{\mu\lambda\nu}\frac{dx^\mu}{ds}\frac{dx^\nu}{ds}\ \xi^\lambda. \tag{102}$$

However, as discussed above, the symplectic structure is due to the fact that Riemann tensor can be derived from covariant derivatives either as

$$\Big[\nabla_\mu, \nabla_\nu\Big] V^\rho = R^\rho_{\sigma\mu\nu} V^\sigma \tag{103}$$

or

$$\Big[\nabla_\mu, \nabla_\nu\Big] V_\rho = R^\sigma_{\mu\nu\rho} V_\sigma \tag{104}$$

In other words, fundamental equations of general relativity are recovered deriving them from symplectic formalism. It is interesting to see that we have never used the metric $g_{\alpha\beta}$ in our considerations, so that our approach is completely affine in Palatini's sense [107]. The interpretation of this fact, leads to the conclusion that to achieve a quantum gravity theory we must search the quantization of connections instead of metric $g_{\alpha\beta}$. An interesting application is obtained keeping the standard 4-momentum of special relativity. In this case, we get the "mass shell condition" [81].

$$\mathcal{H} = p^\alpha p_\alpha = m^2. \tag{105}$$

where, obviously, $p^\alpha p_\alpha = E^2 - \vec{p}^{\,2}$, with E the particle Energy and $\vec{p}$ the momentum. Developing this Hamiltonian formalism for scalar fields, we obtain the results that we will see in detail in section 5, i.e., the Klein-Gordon equation and, after the splitting procedure, the forward and backward Dirac equations [91]. This means that the particular considerations that we are going to develop in section 5, starting from the EPR paradox, can be framed in the general scheme of the covariant symplectic structure and can be generalized in any curved spacetime. Another interesting choice, which allows to recover electromagnetism, start with the vector potential and the Hamiltonian invariant. It is direct, with the same procedure, to obtain the electromagnetic tensor field

$$\mathcal{H} = A^{\alpha} A_{\alpha}. \tag{106}$$

then

$$F_{\alpha\beta} = \nabla_{\alpha} A_{\beta} - \nabla_{\beta} A_{\alpha} = \nabla_{[\alpha} A_{\beta]}, \tag{107}$$

and the Maxwell equations (in a generic empty curved spacetime)

$$\nabla^{\alpha} F_{\alpha\beta} = 0 \qquad \nabla_{[\alpha} F_{\lambda\beta]} = 0. \tag{108}$$

The standard Lorentz gauge is

$$\nabla^{\alpha} A_{\alpha} = 0 \tag{109}$$

and electromagnetic wave equation is easily recovered. It is straightforward to see the formal deep analogy with the Bianchi identities. These examples point out the fact that conservation laws and symplectic structure are strictly related and so let us consider also a quantum invariant whit a vector wave function with α-scalar components. For consistency, it has to be

$$\langle \Psi^{\alpha} | \Psi_{\alpha} \rangle = I \tag{110}$$

where I is the $(\alpha \times \alpha)$ identity matrix. A quantum Hamiltonian invariant can be constructed by inserting covariant derivatives into (110), that is,

$$\langle \hat{\mathcal{H}} \rangle \equiv \langle \Psi^{\alpha} | \nabla_{\alpha} \nabla^{\alpha} | \Psi_{\alpha} \rangle \tag{111}$$

with the standard meaning of the symbols. The quantity $\langle \hat{\mathcal{H}} \rangle = E$ can be interpreted as the Energy of the system. The operator $\nabla_{\alpha} \nabla^{\alpha}$ is nothing else but the d'Alembert operator on a generic (curved) spacetime, so that

$$\hat{\mathcal{H}} \equiv \langle \Psi^{\alpha} | Box | \Psi_{\alpha} \rangle, \tag{112}$$

with the straightforward generalization

$$\langle \hat{\mathcal{H}} \rangle \equiv \langle \Psi^{\alpha} | Box + m^2 | \Psi_{\alpha} \rangle, \tag{113}$$

which gives the Klein-Gordon equation, from which it is easy (considering generic spinors) to obtain forward and backward Dirac equations (see section 5). This means that the following discussion on forward and backward solutions can be immediately reproduced. In fact we get the wave functions

$$|\Psi_{\alpha}\rangle = e^{-i(k_{\alpha}x^{\alpha})}|u(k_{\alpha})\rangle, \qquad \langle \Psi^{\alpha}| = \langle u(k^{\alpha})|e^{i(k^{\alpha}x_{\alpha})} \tag{114}$$

where k_{α} are covariant momenta. However, the argument holds for generic sandwich products $\langle \Phi | \Psi_{\alpha} \rangle$ and functions $f(\langle \Phi | \Psi_{\alpha} \rangle)$. Let's take as a conclusion, that the symplectic-covariant structure is completely general, holds for classical and quantum systems and, above all, it is generated by conservation laws.

4.5. Quantum Cosmology, Physics and Astrophysics. OQR Experimental Confirmations

The aim of Quantum Gravity is to get a coherent scheme in which the gravity is defined with the same formalism of the other interaction [108, 109]. To do that, we have to work at Planck's scale, so the cosmology is the area for the application of the observable predictions for a theory of quantum gravity. In this context, it is supposed that the Universe, as a whole, has a quantum mechanical nature and that a classical universe is only a limit concept valid in particular regions of a manifold (superspace) composed of all the possible space-like 3-*geometries* and local configurations of the matter fields. The task of quantum cosmology is then to relate all the observable quantities of a classical universe to the assigned boundary conditions for a wave function in the superspace.

This wave function should be connected with the probability of obtaining standard universes (even if, in the common approach, it is not a proper probability amplitude since a Hilbert space does not exist in the canonical formulation of quantum gravity) [110]. Quantum cosmology can solve, in principle, the problem of the initial conditions of the classical cosmology: i.e., explain the very peculiar aspect of the universe which we see today, specifying the physical meaning of the boundary conditions for the wave function which describes it, in agreement with the astronomical observations and these conditions have to be compatible with the wave function of the universe $|\Psi\rangle$. The dynamical behavior of $|\Psi\rangle$, in the superspace is described by the Wheeler-DeWitt (WDW) equation [108] that is a second order functional differential equation hard to handle, because it has infinite degrees of freedom. So then it is usual to work on finite dimensional minisuperspace models in which the metrics and the matter fields are restricted to particular forms, like homogeneous and isotropic models. With these choices, the WDW equation becomes a second order partial differential equation which can be, in principle, integrated. A scheme for the solutions of the WDW equation is the Hartle criterion [110], in which he proposed to look for peaks of the wave function of the universe: if it is strongly peaked, we have correlations among the geometrical and matter degrees of freedom; if it is not peaked, correlations are lost. In the first case, the emergence of classical relativistic trajectories (i.e., universes) is expected. In this sense, the boundary conditions on the wave function imply initial conditions for the classical solutions. It is possible to show that the Hartle criterion is always connected to the presence of a conservation law and then to the emergence of classical trajectories [111] which are classical relativistic universes where cosmological observations are possible. Let us now deal with minisuperspace where the wave function of the universe can be consistently defined. The Hamiltonian constraint (corresponding to the $(0, 0)$ Einstein energy equation) gives the WDW equation, so that if $|\Psi\rangle$ is a *state* of the system (i.e., the wave function of the universe), dynamics is given by

$$\mathcal{H}|\Psi >= 0 \tag{115}$$

If conservation laws exist, the reduction procedure, can be applied and then we get

$$\pi_1 \equiv \frac{\partial \mathcal{L}}{\partial Q^1} = ix_1\theta_{\mathcal{L}} = \Sigma_1, \qquad \pi_2 \equiv \frac{\partial \mathcal{L}}{\partial Q^2} = ix_2\theta_{\mathcal{L}} = \Sigma_2, ... \tag{116}$$

depending on the number of symmetries.
After quantization, we get

$$-i\partial_1|\Psi >= \Sigma_1|\Psi >, \qquad -i\partial_2|\Psi >= \Sigma_2|\Psi >, ... \tag{117}$$

which are nothing else but generalized translations along the Q^j axis singled out by the corresponding symmetry (i.e., a conserved physical quantity). Eqs. (117) can be immediately integrated and, being Σ_j real constants, we obtain oscillatory behaviors for $|\Psi\rangle$ in the directions of symmetries, i.e., the universe wave function is given [75] by the superposition

$$|\Psi>=\sum_{j=1}^{m} e^{i\Sigma_j Q^j}|\chi(Q^l)\rangle, \quad m < l \leq n, \tag{118}$$

where m is the number of symmetries, l are the directions where symmetries do not exist, n is the total dimension of minisuperspace. Viceversa, dynamics given by (115) can be reduced by (117) if, and only if, it is possible to define constant conjugate momenta as in (116), that is an oscillatory behavior of a subset of solutions $|\Psi\rangle$ exists only if conservation laws are valid for dynamics. The m symmetries give first integrals of motion and then the possibility to select classical relativistic trajectories. In conclusion, we can set out the following statement: "*In minisuperspace quantum cosmology, the existence of conservation laws yields a reduction procedure of dynamics which allows to find oscillatory behaviors for the general solution of WDW equation. Viceversa, if a subset of solutions of WDW equation has an oscillatory behavior, conserved momenta have to exist and conservation laws are present. If a conservation law exists for every configuration variable, the dynamical system is completely integrable and the general solution of WDW equation is a superposition of oscillatory behaviors. In other words, conservation laws allow and select observable universes*". The major result of the approach is the fact that the quantum cosmology corresponding to a theory where conservation laws are absolutely valid must generate classical universe. The O.Q.R. is a general theory and the resulting scheme is self-consistent and able to solve problems of contemporary physics, because several experimental confirmation comes out from cosmology and astrophysics. For instance the search of the initial conditions [104] of the Universe evolution, in our scheme leads to the relevant result that the existence of conservation laws (i.e., symmetries of the system of the cosmological equations) dinamically determines that the so called "wave function of the Universe" selects first integrals of motion, whose cosmological solutions are "observable universes" [105]). so proposing a solution for an old problem [105, 106]. The OQR approach is confirmed in many cosmological and astrophysical observations, because the model coming out from the reduction process from 5 to 4 dimensions, seems largely fulfill the recent observational results. Up today, combining data from Supernovae of Ia type (SNeIa), from cosmic background radiation (especially COBE, BOOMERANG, WMAP and PLANCK [115, 116, 117, 118]), from galactic clouds Lyα and from the large scale galaxies clusters distribution [108], it comes out a spatially flat cosmological model, whose dynamics is supposed dominated for 30% by matter (barionic and dark) and for 70% by dark energy (which can be considered either cosmological constant, or the so called Quintessence scalar field, or same other unidentified dynamical entity). The characteristic of this observational frame, is that the Universe seems to accelerate, instead to the contrary as in Standard Cosmological predictions, and moreover the possibles sources of dynamics remain totally unknown. With our mechanism, instead, we succeed in reproducing the cosmological parameters in agreement with the observed values, and this starting from a general theory without conditions arbitrarily introduced.

In particular, taking into account our cosmological model [112, 114] derived from the embedding process from 5 to 4 dimensions, model characterized by two parameters, namely the today Hubble constant H_0 and the matter density Ω_M , we get the best fit presented in figure and the final results derived by our singularity free solutions can reproduce Sne Ia data very well.

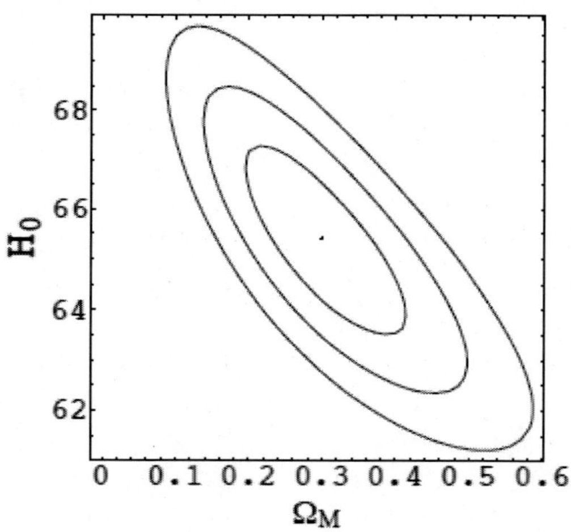

Figure 1. 1, 2 and 3σ confidence region in the (Ω_M, H_0) plane. The central dot represents the best fit: $\Omega_M = 0.3$, $H_0 = 65.4$Km/s/Mpc.

The age of the universe can be directly obtained, by definition, if one knows the value of the Hubble parameter, that is $t_0 = \alpha H_0^{-1}$ where α is a constant depending on the model. For a matter dominated, flat model ($k = 0$), it is $\alpha = 2/3$. We evaluate the age taking into account the above solutions in the 3σ-range of variability of the Hubble parameter deduced by the Supernovae fit and SZE method. We get age estimates included between 13Gyr and 15Gyr. A check for the allowed values of t_0 is to verify if the considered range provides also accelerated expansion rates. We get a negative deceleration parameter, converging to $q_0 \approx -1/2$, for the above, singularity free, solutions evolving toward the future, which is perfectly in agreement with observations. A further test of the model can be performed by the age estimate obtained by the WMAP campaign [117]. Using the above approach, with the only difference to restrict the age estimates between 13.5Gyr and 13.9Gyr, the WMAP results fall perfectly in the ranges theoretically calculated, in the limit of errors. At the end we obtain $\Omega_M \approx 0.3$ (density parameter of barionic and dark matter), $\Omega_\Lambda \approx 0.7$ (density parameter of dark energy), $\Omega_k \approx 0.01$ (spatial curvature), $q_0 \approx -0.5$ (deceleration parameter), $t_0 \approx 14$ Gyrs (estimated age of the Universe), $H_0 \approx 65$ Km / sec / Mpc (Hubble parameter) so reproducing correctly, and at the same time, all the observed values of the five more significant cosmological parameters, with a singularity free, dynamical theory. Another important feature is that the OQR solutions match with different sets of data (e.g. SNeIa and SZE taken at different redshift regimes and epochs) so that the model is consistent from early to late eras of the universe [112, 114]. We should stress that also the scalar field, whose dynamics is ruled by Klein-Gordon equation (i.e., by the Bianchi identities) is recovered in our frame. Another astrophysical problem which can be settled in this scheme, is that of dark matter that should be supposed to explain the observed dynamics of the rotation curves of the galaxies. In fact the effective theory of gravitation, that emerges from the process of dimensional embedding, implies a variable gravitational coupling, in which the Newton constant results to be a parameter dependent from the scale [106]. In other words, the Newtonian potential, able to describe the gravitational effects up to Solar System scale, must be corrected by Yukawa type terms, depending by characteristic lengths, like, for instance, the typical scale of galaxies (from 10 to 100 Kpc). In this scheme, the enormous quantities of dark matter necessary to the astronomers to explain the observed dynamics of the galaxies (up to 90% of the total mass of a typical galaxy) are no longer necessary, because the scale dependence of the gravitational interaction can provide a correct frame, without the addition of elements still unidentified. But also other astrophysical problems can find together a possible solution in the scheme of the Open Quantum Relativity, like Black Holes and Gamma Rays Bursts [104]. The idea is that the absolute conservation of the Mass-Energy, during the gravitational collapse (Black Hole), induces topology changes (at the two junctions of the Worm Hole) in such a way that, in another space-time zone, it emerges an energy fountain (White Hole). This process can induce an explosive emission (peak of the Gamma Rays Burst) with a following tail (spectrum of Gamma Rays Burst). From the calculation it emerges that if the Black Hole is generated by a star big enough (for example of the order of 10 solar masses) it is possible to obtain the energetic values of Gamma Rays Bursts. Moreover, the GRB distribution seems like the distribution of the evolved stars (i.e., stars able to generate Black Holes) and finally the observed polarisations can be naturally explained thanks to the passage of the energetic flux trough the Worm Hole [104]. The black hole and the white hole (in fact a white fountain of GRB) are so two entangled objects, which can live in two different and traditionally considered causally disconnected regions of spacetime. We proposed an OQR approach, (compatible with the Fireball model [65]) able to solve the GRBs puzzle [104] considering them, essentially, as the signature of a process which, to preserve the mass-energy of an evolving black hole, gives rise to a white hole as the ending point of mass-energy flux. The peak of GRBs flux can be connected to the birth of such a structure, while the afterglows are connected to the late and the less energetic story of the system. Our starting point is the key hypothesis that a black hole does not give rise to a "singularity" where the laws of physics do not work, but instead that its mass-energy is conserved in any case and, through a worm hole, dynamically results in a white hole.

Every object falling into black hole (acting as a sink) is so converted in energy emerging from a white fountain into another region of space-time. Black hole and white hole result naturally "entangled" since every perturbation of one of the two systems "necessarily" perturbs the other one without any standard direct interaction between them. In fact the birth of white holes comes from an extremely evolved stars which, exceeding the Chandrasekhar limit, give rise to black holes. The flux peak of GRB is connected to this birth act, while the following luminosity could be connected with the following dynamics around both the edges of black hole and white hole. In OQR approach, the problem of the origin seems solved, because not only the energetics, but also the beaming, the polarization and the cosmological distribution of GRBs can be explained [104]. In fact, the energy flux is naturally beamed by the throat of the worm hole, whose width and length give rise also to the polarization of the burst. The cosmological distribution is due to the fact that this process occurs only in a final stage of star life and the path along the worm hole throat can result in a connection between far regions of the universe. The black holes (due to the late evolution of their progenitor stars) form relatively near to us $(z \sim 0)$, while the white holes emerge far from us at high red shifts. More precisely, the worm hole throat is defined as

$$r = \frac{2G_N M}{c^2} \quad (119)$$

which is nothing else but the Schwarzschild radius. We can define the proper length, which goes to zero as soon as the radial coordinate reaches its minimum. We obtain [104]

$$l_{WH} = \pm\left[\sqrt{r\left(r - \frac{2G_N M}{c^2}\right)} + \frac{2G_N M}{c^2} ln\left(\sqrt{\frac{rc^2}{2G_N M}} + \sqrt{\frac{rc^2}{2G_N M} - 1}\right)\right] \quad (120)$$

The positive component of formula (120) is the proper distance from the black hole while the negative component is the proper distance from the white hole. Such distances directly depend on the mass of the collapsing object. The situation is the following: for $l > 0$, we have the proper distance from the black hole, $l = 0$ implies $r = 2G_N M/c^2$ (i.e., the worm hole throat), $l < 0$ gives the proper distance from the white hole. The proper "length" of the whole system is $2|l|$. The conservation of mass-energy, once the system is stabilized against perturbations [104], is obtained by considering that the mass-energy falling into the black hole, outcomes from the white hole after passing through the worm hole. Let us now show how GRBs can be considered as the signature of Black Hole-Warm Hole-White Hole (BWWH) systems, acting as source-engines. Let us remember that the aspects of GRBs phenomenology, which can be naturally explained by our model, are the energetics, the beaming, the polarization and the fact that GRBs are observed only at high red shifts. Starting with the energetics, black holes are created naturally in the aging of massive stars. Some simple dimensional consideration are necessary now. A solar mass is $M_\odot \simeq 1.9 \times 10^{33} g \simeq 1.1 \times 10^{54}$ Gev, and $1Gev \simeq 1.6 \times 10^{-3}$ erg, so that the mass-energy converted in radiant energy could be of the order of 10^{57} erg, which characterizes GRBs.

It is interesting to note that this is the "natural" energy emission corresponding to the birth act of white holes, where no signal or object can enter but everything has to come out from the event anti-horizon. In other words, the simple request of conservation of mass-energy of a black hole, through a white hole, justifies the huge amount of GRBs emission. Besides, if BWWH system is stable and does not evaporate [104], a part of mass-energy falling into the black hole emerges at the white hole edge as γ-*photons* of wave length of 1.33×10^{-13}

cm, i.e., a γ ray of energy ~ 1 GeV. Moreover, beaming is considered due to the ultra-relativistic motion of source matter, so then the collapse of a massive star gives rise to beamed jets of energy and cannot be isotropic, due to the macroscopic matter flows with Lorentz factors $\Gamma \geq 100$ [126, 128] and polarization is another natural consequence, connected with the achromatic break in the light curve due to the jets produced by the relativistic source [129]. In the case of a BWWH system, we have a longitudinal polarization and a circular polarization due, respectively, to the length and the width of the worm hole throat which generates an extremely anisotropic relativistic beam. Considering the mass-energy involved ($\sim M_{\odot}, 10^{54} erg$) and the relativistic motions, our model fits with the observations [123, 124]. As we will see, the corrections to the Newtonian potential are directly related to astrophysical sizes which, in this case, could be explained without invoking the standard huge amount of dark matter (see also [104]). In fact, in the weak energy limit, fundamental relativistic theories which attempt to unify gravity with the other forces introduce, in addition to the massless graviton, particles with mass which also carry the gravitational force [93, 136]. The general outcome is that a "Geophysical scale window" between the laboratory and the astronomical scales has to be taken into account. Recently, indications of an anomalous long range acceleration, revealed from the data analysis of Pioneer 10/11, Galileo, and Ulysses space-crafts (which are almost outside the Solar System) makes these Yukawa-like corrections come again into the game [133]. Furthermore, Sanders [131] reproduced phenomenologically the flat rotation curves of spiral galaxies with the hypothesis that the additional gravitational interaction is carried by an ultrasoft boson whose range of mass is $m_1 \approx 10^{-27} \div 10^{-28} eV$. The action of this boson becomes significant at galactic scales, without the request of enormous amounts of dark matter to stabilize the systems. Eckhardt [140] uses a combination of two exponential terms and gives a detailed explanation of the kinematics of galaxies and galaxy clusters, again without dark matter models. It is worthwhile noting that both the spacecrafts measurements and galactic rotation curves observational indications come from "outside" the usual Solar System boundaries, used up to now to test General Relativity. However, the above authors do not start from any fundamental theory in order to explain the outcome of Yukawa corrections. In their contexts, these terms are phenomenological while, in the framework of Open Quantum Relativity, we have dynamically derived them from first principles. Another interesting "signature" for Open Quantum Relativity could come from the gravitational waves (GW) detection. Many interferometer and bar detectors are currently in operation, and several interferometers and bars are in a phase of planning [160, 161]. The results of these detectors will have a fundamental impact to probe the General Relativity or every new extended theory of gravitation [162, 163, 164].

A possible target of these experiments is the so called stochastic background of gravitational waves [165, 166, 167, 168]. The production of the primordial part of this stochastic background (relic GWs) is well known in literature [165, 168], where, using the so called adiabatically-amplified zero-point fluctuations process, several authors have shown how the inflationary scenario for the early universe can, in principle, provide the signature for the spectrum of relic GWs. It is shown that linearizing higher order theories of gravity, further gravitational massive modes emerge.

Besides massless spin-2, also spin-0 and spin-2 massive and ghost fields have to be considered. In conclusion, these massive modes could be of interest for direct detection by the LISA experiment. The discovery at the CERN LHC of a candidate of the Higgs boson, the absence of robust signatures of dark matter candidates at fundamental level, the confirmation of Standard Model of Particles, are strong indications that gravitational interaction must be considered in order to get a solution for several puzzles of modern physics ranging from dark energy to mass generation. Summarising, beside the extended theories of Gravity, a straightforward unification result can be achieved enclosing quantum and classical phenomena in the Open Quantum Relativity scheme at infrared and ultraviolet scales [93, 114, 145]. From an astrophysical point of view, OQR do not require finding candidates for dark energy and dark matter at the fundamental level and the approach starts from taking into account only the "observed" ingredients [105]. Detecting new gravitational modes could be a sort of experimentum crucis in order to discriminate among theories, since this fact would be the "signature" that GR should be enlarged or modified. For this purpose, several large gravitational wave (GW) observatories have been operational in the last years up to now: the two 4 km arm length LIGO interferometers in the US [160], the 3 km arm length Virgo in Italy [161], and the 600 m arm length GEO600 [161]. By around 2015, LIGO and Virgo will be upgraded to their so called advanced configurations, and shortly afterwards up to tens of detection events per year are expected. Another planned GW observatory is the Japanese LCGT [161], and the construction of a further large interferometer in Australia is under consideration [160]. Today, the observed expansion rate of the Universe is inconsistent with the prediction of General Relativity (GR) based upon the mass-energy content of the Universe inferred from electromagnetic observations. This means that either GR needs to be modified at large scales or that the Universe contains enormous amount of (dark) mass-energy, that is not visible in electromagnetic observations. Assuming the possibility that effective gravity could contain further degrees of freedom [93, 177], let us generalize the action of GR by adding curvature invariants other than the standard Ricci scalar. Living to the quoted paper [179] for the full treatment, the contributions to the gravitational radiation, coming from higher order gravity, could be efficiently selected if it would be possible to investigate gravitational sources in extremely strong field regimes. In such a case, the further polarizations coming from the higher order contributions could be, in principle, investigated by the response of a GW detector. However, this situation seems futuristic at the moment, so the realistic approach, to investigate these further ontributions, seems the cosmological background, in particular the stochastic background of GWs. Our study covers extended gravity models with a generic class of Lagrangian densities with higher order terms of the form $f(R, P, Q)$, where $P \equiv R_{ab}R^{ab}$ and $Q_{abcd}R^{abcd}$.

We have linearized the field equations for this class of theories around a Minkowski background and found that, besides a massless spin-2 field, the theory contains also spin-0 and spin-2 massive modes with the latter being, in general, ghosts. Then, we have investigated the detectability of additional polarization modes of a stochastic GW with ground-based laser-interferometric detectors and space interferometers. Such polarization modes, in general, appear in ETGs and can be used to constrain theories alternative to GR in a model independent way. For the situation considered here, we find that the massive modes are certainly of interest for direct attempts at detection.

In conclusion, the approach presented here widely enlarge the possibility of detecting gravitational waves since the whole budget of the gravitational degrees of freedom has be considered. Among the characteristics of OQR (which is an extended theory of Gravity) it is very interesting to point out that the Yukawa-like corrections are naturally emerging in the weak field limit, giving rise to a characteristic length, connected to massive graviton modes, that is capable to address, in principle, the sizes of astrophysical self-gravitating structures [180]. In particular, upper limits on graviton mass well agree with characteristic sizes of galaxies. The issue of the graviton mass has potentially important cosmological implications, e.g. in connection with attempts of modifying Einstein gravity at cosmological distances (which is largely motivated by the problem of dark matter and dark energy). Finally, we want just mention a recent result is the correct reproduction of the observational limits of neutrino oscillations, derived from induced gravitational field in our theoretical scheme [176].At the end, the whole subject presented in this paragraph is extremely interesting from a theoretical point of view, because it means that the puzzle of why masses and sizes of self-gravitating structures range in well defined intervals, could be addressed in a pure gravitational way, without invoking huge amounts of dark matter necessary if the gravitational potential is the standard Newtonian one as pointed out by GR.

5. Time and Time Travel Possibility in Open Quantum Relativity

We are going now to deal with Time in OQR, theory starting from the main stream of standard General Relativity and Quantum Mechanics, but providing another point of view, a general synthesis, in the light of the General Conservation Principle [90]. The first characteristic of this new scheme is that it is not violating Relativity principles but it is extending their validity to include backward time solutions as a "necessity" to preserve always the conservation laws and the causality principle. The Many Worlds interpretation of Quantum Mechanics [1] is another necessity of this approach, coming out as a final point of a process which starts from the very basic principles and ends with very general results. Namely: the solution of the EPR paradox as a conflict between Relativity and Quantum Mechanics [86]; the generalization of entanglement concept as a gate, through topology changes, to recover causality thanks to the backward causation [85]; and finally a dynamical scheme for the unification of the different interactions [93]. We must underline that the whole question of time, and of time machine, is completely changed by this new picture, because "moving in time" is no longer an extreme possibility, only due to the relative motion of two frameworks on the same time arrow, but a more general possibility related to the existence of a second and backward directed time arrow [88]. In the "conventional" time machine, backward in time means only a relative or circular past, while, in the new scheme, there is also another more general possibility and the real question is how to get "physically" a backward time arrow, that we cannot ordinarily perceive, but which has to exist [85, 86, 88]. In fact, as we will see in the next paragraph (5.1), the equations (142) describing quantum matter, can be interpreted respectively, as progressive and regressive solutions in four combination. It comes out that a function (a superposition) of the form $\varphi(x) = \alpha_1\psi(x) + \alpha_2\psi^*(x)$, is a general solution of the dynamics and the states ψ and ψ^* can be interpreted as entangled since they can influence each other also when they are considered disconnected. This point, is of general meaning, because it is suggesting that the existence of two time arrows is a necessary feature of Nature, independently of our capability of perceiving them. The necessity to take into account two arrows of time is implying the consideration of a fifth dimension to open both directions in time and moreover, in this picture, the fifth dimension is not a mathematical tool, but a feature of real physical meaning, because, from the embedding of this fifth dimension, it has been shown that is possible to get mass, charge and spin of all particles [145]. The resulting scheme seems properly to work in recovering basic principle, like conservation and causality, with the minimal necessary number of parameters, in comparison to all existing theories (like Strings, or Supergravity). Particularly interesting, is the fact that entangled gravitational systems constituted by black holes, wormholes and white holes naturally emerge, through topology changes, starting from the request that the mass-energy of collapsing systems is conserved. The main point is that such systems can be stable so that (due to conservations avoiding symmetry breakings) time travels and time machine become, on a theoretical ground, a real possibility [104], and moreover, starting from 5D dynamics, CTCs are ordinary solutions of field equations.

Due to that, entangled gravitational systems (in 4D) are a straightforward necessity of the theory [85, 104], and we will see that, from the reduction mechanism, we get wave like solutions for the Klein Gordon with intrinsecally two times and, because of the structure of the functions $f(t, x_j)$, closed time-like paths are naturally allowed possibilities starting from 5D-dynamics. While the time travels in G.R. are related only to the relativistic effects among frameworks in relative motion on the same time arrow, here, in this approach, are the natural result of an underlying feature of Nature [93, 88]. This feature comes out every time a topology change is necessary, so when and if a conservation law would be otherwise violated. Even if starting from Q.M. considerations, there is the full recovering of Relativity (because we do not need to travel faster than light) but the mechanism of relative motion between two frameworks, which is generating the hypothesis of time travels, is clearly not the same if applied on one or two time arrows. In this sense, we speak in terms of Open Quantum Relativity [88].

5.1. Two Time Arrows as General Feature of Nature the EPR Paradox

Let us now show, following our approach, in which way the two times arrow formally comes out. In a famous paper (EPR Paradox [67]), Einstein, Podolsky and Rosen stated that Quantum Mechanics, violating the locality principle, brings to results which contradict the Relativity. In fact, the foundations of Quantum Mechanics, mainly in relations to Relativity, are always largely discussed, since the intrinsic characteristic of Quantum Mechanics is the existence of systems that have not definite values of measurable quantities, unless one does not measure them. The fundamental point is that this comes from the fact that the state of a system is a superposition of different states, and the only possibility of observation consists in interacting in an irreversible way with the system, by the measure process. This is described as the collapse of the wave function by the measure, and the evolution of the system, after the measure, has the collapsed state as initial wave function. This description, mainly due to Bohr, is known as Copenhagen Interpretation (CI), and following it, the irreversibility of the measure process (and consequently the irreversibility of the time direction), plays a fundamental role. Moreover the superposition of quantum states, that in CI is characteristic of the microscopic description of Nature, can have macroscopic effects and this is at the base of the EPR paradox. Let us take into account (looking at the Bohm's clear example [80]) a spin zero particle, or in general a bound state, which decays into two particles each of spin 1/2.

As far as the spin of the particles in a definite direction is concerned, the state of the system is described by a state vector of the following form:

$$|0\rangle = |\uparrow\rangle|\downarrow\rangle - |\downarrow\rangle|\uparrow\rangle, \tag{121}$$

where we remember that every single particle is not a state by itself and the evolution of the state vector, with respect to the spatial distribution, of the wave function is not yet specified. We know that at the time $t = t_0 = 0$, in which the decay occurs, the spatial support of the variable $\overrightarrow{x}_1 \equiv \{x_{1,1}, x_{2,1}, x_{3,1}\}$, (position of the particle 1) which is in the wave function $\psi(\overrightarrow{x}_1, \overrightarrow{x}_2, \sigma_1, \sigma_2, t_0)$ (with σ_1 and σ_2 spin of the particles) coincides with the spatial support of the variable $\overrightarrow{x}_2$ (position of the particle 2). Now we let the system evolve with an Hamiltonian $\mathcal{H}$ such that, after a time $\Delta t = t_m - t_0$ (where t_m is the time of the measurement), the wave function of the system

$$\psi(\overrightarrow{x}_1, \overrightarrow{x}_2, \sigma_1, \sigma_2, t_m) = e^{\frac{i}{\hbar}\mathcal{H}\Delta t}\psi(\overrightarrow{x}_1, \overrightarrow{x}_2, \sigma_1, \sigma_2, t_0) \tag{122}$$

is such that the spatial supports in x_1 and x_2 are separated, i.e., the particles coming from the decay get their individuality. Note that, in the CI, the evolution of the system after the measure is given by (if, for example, the measure of σ_1 is $+1/2$)

$$\psi(\vec{x}_1, \vec{x}_2, 1/2, -1/2, t) = e^{\frac{i}{\hbar}\mathcal{H}(t-t_m)}\psi(\vec{x}_1, \vec{x}_2, 1/2, -1/2, t_m) \quad (123)$$

Note that we do not perform any measurement at time t, being $t_0 < t < t_m$ and therefore we do not know anything about the state of the system in that range. Now, at time t_m, we measure the spin σ_1 of the particle 1. The spin state of the system is still determined by Eq. (121), but now the measure operation causes the collapse of the wave function in one of the two states $|\uparrow\rangle|\downarrow\rangle$ or $|\downarrow\rangle|\uparrow\rangle$. This implies that, once the spin of particle 1 is measured, also the spin of particle 2 is instantaneously acquired, although the particles are far apart. Therefore it comes out the following paradox : we have two particles in absence of direct interaction, but the state of the system is such that, if we fix the spin state of one particle, also the spin state of the other particle is instantaneously fixed. So, at least apparently, contradicting the limit of the (finite) light speed as maximum possible speed value and after all the whole Relativity. The non local behavior (connected with the Bell's inequalities [18]) of this kind of systems has been tested by Aspect *et al.*, [81] and the experimental results show that the non locality is an actual feature of Nature. The experiments were performed with a slightly different device, i.e., the products of the decay were photons, but, even if this is a particular case, the formal description of the experiment remains the one described above. Let us reasume the results of this experiment (and of the many others made by different groups of physicists) using Sakurai's words (in a book revised by J. Bell himself [17]): "All the experiments made, have conclusively shown that Bell's inequality (which comes from the "locality prescription" of Einstein) is violated and violated in a way which is compatible, within the errors, with Quantum Mechanic's prediction". In order to settle this paradox, very crucial because at the intersection between Quantum Mechanics and Relativity, we developed a new approach, starting from the remark that the Noether Theorem [82] states that for every conservation law of Nature a symmetry must exist. From this statement, it comes out directly (as we will see sun) the consideration of a backward time evolution of wave function, since dynamics, if derived from a variational principle, is always symmetric under time reversal transformations.

This general remark implies that conservation laws intrinsically contain forward and backward causation, even if against common sense and local realism, which assume just one time arrow. We are going to show, by very general arguments, that Bianchi's identities, which are geometric identities directly connected to conservation laws, contain symmetric dynamics. From such a dynamics, it is possible to describe backward and forward evolution of the wave function, considering that quantum matter can be described by a scalar field ϕ [83, 84]. As phenomenological definition, the entanglement takes place when two or more physical objects, despite being spatially disconnected, are subjected to an inter-relation for which the effect of a perturbation on one of them induces a perturbation on the other one, without any direct known interaction on each other. In previous paper [85, 86], we proposed a solution of the EPR paradox, which is preserving all the features of CI, but provides a dynamics and a generalisation, thank to the fact that the time flow of the wave function can be different from the time flow of the observer. As we outline, this possibility, although not intuitive and not referable to our perception of reality, is compatible with Quantum Mechanics, since our basic ignorance of the system state in the time interval $t_0 < t < t_m$. The dynamics of a scalar field ϕ, describing quantum matter on a (curved) space-time, is given by the stress-energy tensor

$$T_{\mu\nu} = \nabla_\mu\phi\nabla_\nu\phi - \frac{1}{2}g_{\mu\nu}\nabla_\mu\phi\nabla^\mu\phi + g_{\mu\nu}V(\phi), \tag{124}$$

which is a completely symmetric object, being

$$T_{[\mu,\nu]} = T_{\mu\nu} - T_{\nu\mu} = 0. \tag{125}$$

$V(\phi)$ is a self-interacting potential specified below; ∇_μ is the covariant derivative on a given spacetime manifold (as standard in General Relativity, we assume a Riemannian manifold with a metric tensor $g_{\mu\nu}$ defined on it); the indexes are $\mu, \nu = 0, 1, 2, 3$ and the signature of the metric is $(+ - --)$. Since (ϕ) is a scalar field, the covariant derivative coincides with the partial derivative so that

$$\nabla_\mu\phi = \partial_\mu\phi = \phi_\mu. \tag{126}$$

Such a tensor has to satisfy the conservation laws

$$\nabla_\mu T^\mu_\nu = 0, \tag{127}$$

which are,at the end, the contracted Bianchi identities for $T_{\mu\nu}$. Sending to [86] for the full derivation, it is straightforward to show that

$$\nabla_\mu T^\mu_\nu = \phi_\nu \Box\phi + V_\nu = \phi_\nu \left[\Box\phi + \frac{dV}{d\phi}\right], \tag{128}$$

where $\Box$ is the d'Alembert operator and $\nabla_\mu \phi^\mu = \Box\phi$. The result is that the validity of the contracted Bianchi identities, implies the Klein-Gordon equation which gives the dynamics of ϕ, that is, obviously assuming $\phi_\nu \neq 0$,

$$\nabla_\mu T^\mu_\nu = 0 \quad \Longleftrightarrow \quad \Box\phi + \frac{dV}{d\phi} = 0 \tag{129}$$

It is interesting the full symmetry of the result, i.e., the Klein-Gordon operator is symmetric. Specifying the problem to the case of a self-interacting massive particle, we can write

$$V(\phi) = \frac{1}{2}m^2\phi^2, \quad \textit{and then} \quad \frac{dV}{d\phi} = m^2\phi, \tag{130}$$

so that we obtain the Klein-Gordon equation

$$(\Box + m^2)\phi = 0. \tag{131}$$

Being ϕ the scalar field, we want to stress that it can be interpreted as the product of two conjugate complex Ψ^*, Ψ

$$\phi = \Psi^*\Psi \tag{132}$$

Now, let us consider, as an example, the case of a Minkowski space-time, and develop the Klein-Gordon operator (suppressing the indexes)

$$\Box + m^2 = \partial_\alpha\partial^\alpha + m^2 = (\partial - im)(\partial + im) \tag{133}$$

For consistency, the Klein-Gordon equation gives

$$(\Box + m^2)\phi = (\partial_\alpha\partial^\alpha + m^2)(\psi^*\psi) = (\partial - im)(\partial + im)(\psi^*\psi) = 0 \tag{134}$$

and Eq. (134) can be split, for massive particles, in the cases

$$(\partial - im)\Psi = 0 \qquad (\partial + im)\Psi^* = 0 \tag{135}$$

$$(\partial - im)\Psi^* = 0 \qquad (\partial + im)\Psi = 0 \tag{136}$$

Instead, for massless particles, the four conditions reduce to two

$$\partial\psi = \partial\psi^* = 0. \tag{137}$$

Taking into account spinors, we can write down

$$(i\gamma^\mu\partial_\mu - m)\psi = 0, \tag{138}$$

and analogous equations for the other cases, where γ^μ are the standard [82] Dirac matrices

$$\gamma^0 = \begin{pmatrix} I & 0 \\ 0 & -I \end{pmatrix},$$

$$\gamma^i = \begin{pmatrix} 0 & \sigma^i \\ -\sigma^i & 0 \end{pmatrix}, \tag{139}$$

with I the $(2x2)$ identity matrix and σ^i the three Pauli matrices [83]. We can see that Eqs.(135) can be considered as a "forward" and "backward" propagators and viceversa for the other two (136). In terms of 4-momenta, we have

$$(\partial - im) \quad \longrightarrow \quad (k - im), \tag{140}$$

$$(\partial + im) \quad \longrightarrow \quad (k + im), \tag{141}$$

so the general solutions have the forms

$$\psi(x) = e^{-ikx}u(k), \qquad \psi^*(x) = e^{ikx}u^*(k) \tag{142}$$

which can be interpreted, respectively, as progressive and regressive solutions. It comes out that a function (a superposition) of the form

$$\varphi = \alpha_1\psi + \alpha_2\psi^*, \tag{143}$$

where $\alpha_{1,2}$ are arbitrary constants, is a general solution of the dynamics and the states ψ and ψ^* can be considered, in our scheme, as entangled since they can influence each other also when they are disconnected. At the end, the absolute validity of conservation laws gives rise to a symmetric dynamics (backward and forward evolution of the system) and the entanglement of states is naturally determined. We want to stress that we found the four conditions (135)-(136) which satisfy Eq.(134), and this implies that backward and forward evolutions exist both for the ψ-field and the conjugate ψ^*-field. In some sense, it seems that all the folds of light-cone, in Minkowski spacetime, have the same dignity, but we have been confined to see (at least macroscopically) just the fold toward the future (the arrow of time which we can usually perceive). Finally, by writing the wave function in polar representation

$$\psi = e^{-i\theta}\rho \tag{144}$$

it is immediate to see that backward or forward evolution depends on the sign of the phase θ, and then a pure geometric representation is allowed for time evolution. As a conclusion, in order to explain the fact that a measurement made on a particle seems able to affect the state of another particle, entangled but spatially disconnected, it is important to notice that the statement can be reexpressed in terms of an effect of a measurement of a particle able to affect the system in the past: i.e., the relation of interference among variables, known as Bell inequalities, can be obtained if the particles decay in a state that depend on what will be measured [86]. The above considerations are completely general and can be developed in the framework of curved space-times, as we discussed in the framework of the covariant symplectic structure. The starting point of the new way to see at the time travels in OQR is exactly here.

5.2. Causality, Entanglement and Topology Changes

The EPR paradox, apparently showing the incompatibility between Causality and Quantum Mechanics, leads to a deep crisis of the concept of Causality as a Principle of general validity. This fundamental question has become even more urgent, because, after the Bell's Inequalities and the related experiments [81], Quantum Mechanics is confirmed as a description fundamentally correct, so leaving until now the problem unresolved, despite many attempts and exotic hypotheses like the "hidden variables", by definition impossible to be revealed and so then epistemologically unsatisfactory. In OQR, the redefinition of the entanglement concept, based on the generalization to a five dimensional space, leads instead to the recovery of the classical and fundamental idea of Causality.

The absolute validity of the conservation laws induces, in the cases where they would be otherwise violated, topology changes which make possible "tunnels" (like wormholes in relativistic astrophysics) connecting separated space-time regions. This phenomenon turns out to be describable as a-luminal and able to allow, without overcoming the light speed limit, the existence of conceivable time-tunnels. In fact, Quantum Mechanics states that an interaction, even if acting only on a single part of a quantum system, determines a dependent evolution of the correlated quantities of the other part, also when these parts are placed in regions "causally disconnected" of spacetime, (i.e., when no direct interaction between these parts can occur, in the light speed limit). From the beginning, Schröedinger tried to overcome the problem by the qualitative concept of entanglement, defined as a sort of deep connection, not yet dynamically specified, able to link two causally disconnected, but quantum related, objects. Starting from this qualitative concept, later accepted and elaborated by many authors [97], in a previous paper [85], we recovered the causality giving, in the light of OQR, this new entanglement definition: "*Two states, spatially separated and causally disconnected in the four dimensional space that we ordinarily perceive, are entangled if an interaction with one of them can influence the other one, without in any way directly interact with it, because a four dimensional entanglement means that it exists a causality nexus in a larger five dimensional physics space*". The reason leading to this definition is a real one, since the entanglement, unavoidable concept to explain why Quantum Mechanics works, would be otherwise simply impossible, without violating the causality principle and also the pure logic, which clearly states that: "It is impossible an interaction with an object without in some way interacting with it". If we do not hypothesize another physical dimension (the fifth one), in which the two states are causally connected, in that way restoring the causality principle, what we call entanglement would remain a "necessary but inexplicable" phenomenon. In our theory, the conservation laws are the first principle which determines all the following evolution, since the fact that they can never be violated brings to a mechanism avoiding such a violation, also in the cases in which, for the standard interpretations, the violation should occur. This mechanism is the topology change. The Entanglement so maintains the meaning of the underlying mechanism allowing interactions otherwise impossible (which initially Schröedinger gave to it), but with an endowed dynamics providing an explanation of the phenomenon which emerges in 4D-spacetime thanks to the existence of a larger dimensional space. Now, conservation laws are defined in Q.M. through the Heisenberg relations

$$[\Sigma_j, \mathcal{H}] = 0. \tag{145}$$

With the meaning that j momenta of the n-dimensional phase space are conserved. Clearly, a conservation law is not admitted if

$$i\frac{\partial\Sigma_j}{\partial t} = [\Sigma_j, \mathcal{H}] \neq 0, \tag{146}$$

where we are assuming Planck units with $\hbar = c = k_B = 1$. With π_j conjugate momenta, by the Dirac canonical quantization procedure, we have by definition

$$\pi_j \longrightarrow (\hat{\pi}_j = -i\partial_j), \quad \mathcal{H} \longrightarrow \hat{\mathcal{H}}(q^j, -i\partial_{q^j}), \tag{147}$$

If $|\Psi\rangle$ is a state of the system (i.e., the wave function of a particle), dynamics is given by the Schröedinger eigenvalue equation

$$\hat{\mathcal{H}}|\Psi\rangle = E|\Psi\rangle \tag{148}$$

where the whole wave-function is given by $|\varphi^{(t,x)}\rangle = e^{iEt/\hbar}|\Psi\rangle$. If conserved quantities exist, we get classically

$$\pi_1 \equiv \frac{\partial\mathcal{L}}{\partial\dot{Q}^1} = i_{X_1}\theta_{\mathcal{L}} = \Sigma_1, \quad \pi_2 \equiv \frac{\partial\mathcal{L}}{\partial\dot{Q}^2} = i_{X_2}\theta_{\mathcal{L}} = \Sigma_2, \; .... \; \pi_j \equiv ..., \tag{149}$$

where $\mathcal{L}$ is a canonical Lagrangian, i_{X_j} is a contraction and $\theta_{\mathcal{L}}$ is the Cartan one-form, the index j depending on the number of symmetry vectors. After Dirac quantization, we get

$$-i\partial_1|\Psi\rangle = \Sigma_1|\Psi\rangle, \quad -i\partial_2|\Psi\rangle = \Sigma_2|\Psi\rangle, ... \quad -i\partial_j|\Psi\rangle = \Sigma_j|\Psi\rangle, \tag{150}$$

which are nothing else but translations along the $\mathcal{Q}^j$ axis singled out by the corresponding conserved quantity. Eqs. (150) can be integrated and, being Σ_j real constants, we obtain oscillatory behaviors for $|\Psi\rangle$ in the directions of such quantities, i.e.,

$$|\Psi\rangle = \sum_{j=1}^{m} e^{i\Sigma_j Q^j}|\chi(\mathcal{Q}^l)\rangle, \quad m < l \leq n, \tag{151}$$

where m is the number of symmetries, l are the directions where symmetries do not exist and n is the number of dimensions of configuration space. Vice-versa, dynamics given by (148) can be reduced only if it is possible to define constant conjugate momenta as in (149), i.e., the oscillatory behaviors of a subset of solutions $|\Psi\rangle$ exist as a consequence of the fact that conserved quantities are present in the dynamics. The m symmetries give first integrals of motion so then the possibility to select exact solutions for particles and, moreover, if $m = n$ the problem is completely solvable and a symmetry exists for every variable of configuration space. At the end, we can say that the existence of conservation laws, determines also the structure of the configuration space (a vector space in the case of Hilbert) where the physical system is set. Vice-versa, the degree of solvability and separability of a system is deeply related to the existence of conservation laws and, moreover, all physical quantities are conserved in a completely integrable system. So then in Quantum Mechanics the Eq. (145) assumes the meaning of Bianchi identities and we can see that entangled states like

$$|\Psi\rangle = \frac{1}{\sqrt{2}}\{|\chi\rangle_1|\eta\rangle_2 + |\chi\rangle_2|\eta\rangle_1\} \tag{152}$$

rise up when the superposition (151) is not possible, i.e., when the Heisenberg relations (145) do not hold. Such a mechanism of entanglement is necessary in a 4D-spacetime, but not in a 5D-one, where the Hamiltonian is always conserved. At the end, the entanglement of quantum states comes from the necessity to overcame the impossibility to write down the wave function of a system as a superposition.

In the light of OQR, it finds a possible solution also the question of the hypothesized contemporaneousness of the effects of an interaction on one part of a system, with the induced effect on another part (placed far and non directly interacting, but entangled). This question has become even more important after recent results on quantum teleportation [77, 78], where it appear possible transfer informations instantaneously between separated, but quantum related, regions. This would be a major problem, since in clear contradiction with the basis of Relativity itself, if we continue to treat it in terms of super-luminality, while it is no longer a puzzle if treated in terms of a-luminality. Looking at Fig. 2, the question finds a possible solution. Let us take into account a bounded surface: it is evident that it would take time to go from a point at the upper side to the corresponding image point at the lower side of such a surface, but this time is reduced to zero, if a mechanism exists to make an hole and get directly the bottom from the top. It is straightforward to see that, once defined a standard Lorentz transformation for time and space intervals

$$\Delta t' = \Delta t(1 - v^2/c^2)^{1/2}, \qquad \Delta x' = \Delta x(1 - v^2/c^2)^{1/2}, \tag{153}$$

the reduction to zero of the space interval $\Delta x'$ implies, in the second of (153), that $v = c$, so then the time interval is $\Delta t' = 0$. The situation is described in Fig. 2: the travel is from A to B (or viceversa), the mechanism is the topology change and the deep reason of this is the physical necessity to save the conservation laws, without overcoming the light speed. This type of travel is a-luminal.

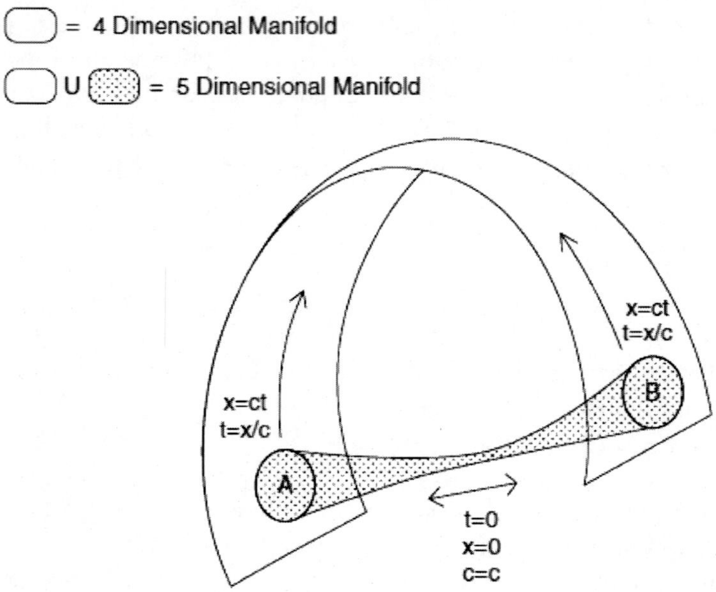

Figure 2. A graphic representation of an a-luminal travel involving a Topology Change.

The fact that conservation laws can never be violated leads, in case of necessity, to a topology change. It is the topology change, which provides the dynamics allowing the particular interaction between two otherwise causally disconnected states that we call entanglement, it is the topology change which makes possible the "hole" connecting the top and the bottom of the example of Fig. 2 or, more specifically, the formation of a wormhole connecting a black hole with its entangled white hole, via a Goedel type change of topology [93, 97]. The fundamental concept which has to be stressed here, is the "a-luminality", since it saves conservation laws and causality principle, and it seems moreover the only mechanism capable to be compatible with the necessity to extend, but not contradict, the Relativity. The new results of the contemporary physics, are impossible to be explained with the standard formulations of Quantum Mechanics (QM) and General relativity (GR), without destroying traditional well founded principles like the causality principle (impossible to recover in the standard Copenhagen interpretation) or conservation laws. However, unifying at the very beginning QM and GR, instead to try to quantize GR a posteriori, leads to a recovering of both causality principle and conservation laws, giving a new meaning to the entropy and to the relation between information and observers. Another interesting opportunity, is the possibility to investigate a quantum universe, starting from a minimal set of variables, defining the quantum state in the framework of Open Quantum Relativity. The way to link different regions of the universe, even if (apparently) causally disconnected, as described by cosmological quantum states, is the entanglement between the states. As a consequence a model emerges from this picture, giving us, in this way, the possibility to realize a statistical and quantum interpretation of the cosmological constant. In particular it appears possible that a correlation of the density of Λ, Ω_Λ, from the mass and curvature densities (i.e., respectively Ω_m and Ω_k) is easily found, by considering the grade of mixed states, while the entanglement plays the role of the physical process allowing this mixing. As a consequence, the concept of causality results enlarged (and recovered) and the entanglement has a physical role for the construction of a quantum state of the universe (which is impossible considering only the standard superimposable states) and finally all the universe is correlated by this process, recovering the causality principle in a generalized way [67]. Furthermore, it is possible to infer a cosmological state, by starting from mixed states so then we will be able also to describe a correlation between states which allows to have an expression for Ω_Λ. We refer to mixed states, dealing with the possibility to have states which behaves as $|\Psi\rangle \approx |\phi_1\rangle|\phi_2\rangle$. Hence, analysing the mixed state ansatz, we notice that, in order to describe a functional dependence of Ω_Λ from $\Omega_{m,k}$, for a given era, entanglement becomes not necessary, and it will be sufficient mixed states only. Starting from, the density matrix for the mixed states, we can arrive (see paper [111] for the full treatment) to write a form of entropy for universe as follows

$$S = -\lambda_- ln\lambda_- - \lambda_+ ln\lambda_+, \tag{154}$$

which derives from the definition [83] of the Von Neumann entropy $S = -Tr(\hat{\rho} ln\hat{\rho})$ Mixed states are physical observables by which it is possible to build up the functional dependence $\Omega\Lambda = \Omega\Lambda(\Omega_{m,k})$ but if we need an universe in which all the regions are connected by a causality principle, we must have a phenomenon of entanglement.

The basic use of the entropy S, inherent to an entangled state, is the role of measurement of information and of correlation between two regions of universe. So the thermal entropy would really derive from the quantum expression of Von Neumann entropy, inherent to the quantum picture proposed. In the simplest ansatz we adopted (two regions of different space-time epoches) we can note that, in general, the Von Neuman entropy obeys to the inequality

$$S(p) \leq S(\rho^A) + S(\rho^B) \tag{155}$$

where $\rho = \rho^{AB}$. From that property of S, it is possible to find the so-called mutual information [83] as follows:

$$I(\hat{\rho}^A || \hat{\rho}^B) = S(\hat{\rho}^A) + S(\hat{\rho}^B) - S(\hat{\rho}) \tag{156}$$

This relation represents the superposition of the two regions entropies, that is to say that we have to imagine that an universe composed by two eras, or regions, each of them having an its own entropy value, identified by the reduced density matrix $\hat{\rho}^A$, for first era and, otherwise, by $\hat{\rho}^B$, for the other one. However, this scheme can be extended to any number or regions.

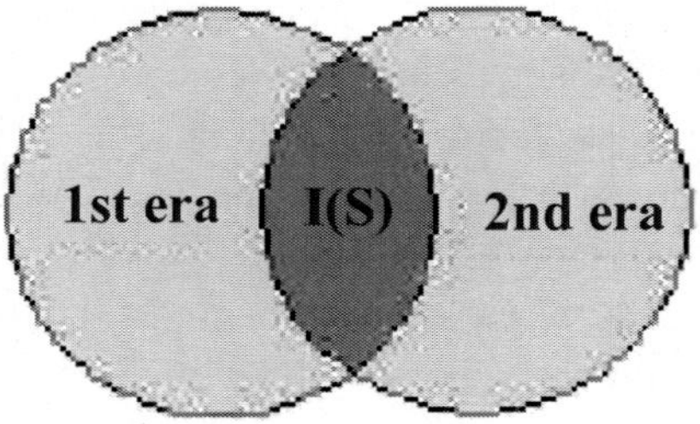

Figure 3. A scheme for understanding the role of mutual information $I(S)$, in the context of two quantum regions. The degree of superposition of two regions is directly proportional to $I(S)$.

Entanglement suggests how to construct a quantum state for the universe using the theoretical tools of Open Quantum Relativity, i.e., considering a quantization procedure at the very foundation of the theory. In this way, GR is recovered from Entanglement and it is described at the end of the formulation of

Ω_Λ in order to understand the role of Λ and of other objects in cosmology. This generality is a peculiar feature of Open Quantum Relativity [145], thanks to the fact that it generalizes the causality principle (including entanglement). The question if a physical system can be considered existing either with or without information is solved in the sense that the physical system without information is not interacting with systems outside of it. This means that the solutions with real values are what we are ordinarily experiencing while the immaginary solutions are not producing observable physical effects unless we connect them by entanglement. The emergence of "entropy", in the usual sense, suggests to take into account only one time arrow until we do not take into account imaginary solutions and the entanglement phenomenon. In such a case a backward dynamics (and then two time arrows) have to be considered.

5.3. Scalar Tensor Gravity from 4D Reduction Two Time Arrows & Closed Time-like Paths

The general consideration in 5D, can be reduced to a 4D-dynamics taking into account the Campbell theorem [95]. This theorem states that it is always possible to consider a 4D Riemannian manifold, defined by the line element $ds^2 = g_{\alpha\beta}dx^\alpha dx^\beta$, in a 5D one with $dS^2 = g_{AB}dx^A dx^B$. We have $g_{AB} = g_{AB}(x^\alpha, x^4)$ with x^4 yet unspecified extra coordinate and the metric g_{AB} is covariant under the group of 5D coordinate transformations $x^A \rightarrow \bar{x}^A(x^B)$, but not under the (restricted) group of 4D transformations $x^\alpha \rightarrow \bar{x}^\alpha(x^\beta)$. This relevant fact has, as a consequence, that the choice of 5D coordinates results as the gauge necessary to specify the 4D physics also in non-standard aspects (e.g. entanglement [85]). Vice-versa, in specifying the 4D physics, the bijective embedding process in 5D gives physical meaning to the fifth coordinate x^4 [93]. Moreover, the signature of the fifth coordinate results from the dynamics generated by the physical quantities which we observe in 4D (mass, spin, charge) and such a process gives rise to fundamental features. Leaving to the quoted paper [75] for the full demonstration, let us start our considerations by the most general action in 4D in which gravity is nominimally coupled to a scalar field:

$$\mathcal{A} = \int_{\mathcal{M}} d^4x\sqrt{-g}\left[F(\phi)R + \frac{1}{2}g^{\mu\nu}\phi_{;\mu}\phi_{;\nu} - V(\phi) + \mathcal{L}_m\right] + \int_{\partial\mathcal{M}} d^3x\sqrt{-h}K \quad (157)$$

(Basini-Capozziello Action [93]) where the form and the role of $V(\phi)$ are still general and $\mathcal{L}_m$ represents the standard fluid matter content of the theory generated by the splitting of the 5D Klein-Gordon equation. The state equation of fluid matter is $p = \gamma\rho$ and $0 \leq \gamma \leq 1$ where p and ρ are, respectively, the ordinary pressure and density. The second integral is the boundary term where $K \equiv h^{ij}K_{ij}$ is the trace of the extrinsic curvature tensor K_{ij} of the hypersurface $\partial\mathcal{M}$ which is embedded in the 4D manifold $\mathcal{M}$, and h is the metric determinant of the 3D-manifold. The field equations can be derived by varying with respect to the 4D-metric $g_{\mu\nu}$

$$G_{\mu\nu} = \tilde{T}_{\mu\nu}, \tag{158}$$

where

$$G_{\mu\nu} = R_{\mu\nu} - \frac{1}{2}\, g_{\mu\nu}R \tag{159}$$

is the Einstein tensor, while

$$\tilde{T}_{\mu\nu} = \frac{1}{F(\phi)}\Big\{-\frac{1}{2}\phi_{;\mu}\phi_{;\nu}+\frac{1}{4}g_{\mu\nu}\phi_{;\alpha}\phi^{;\alpha}-\frac{1}{2}g_{\mu\nu}V(\phi)-g_{\mu\nu}\Box F(\phi)+F(\phi)_{;\mu\nu}+T^{m}_{\mu\nu}\Big\} \tag{160}$$

is the *effective stress-energy tensor* containing the nominimal coupling contributions, the kinetic terms, the potential of the scalar field ϕ, and the ordinary matter stress-energy tensor.

In the case in which $F(\phi)$ is a constant F_0 (in our units, $F_0 = -1/(16\pi Gn)$) we get the usual stress-energy tensor of a scalar field, minimally coupled to gravity, that is

$$T^{\phi}_{\mu\nu} = \phi_{;\mu}\phi_{;\nu} - \frac{1}{2}g_{\mu\nu}\phi_{;\alpha}\phi^{;\alpha} + g_{\mu\nu}V(\phi) \tag{161}$$

By varying with respect to ϕ, we get the 4D-Klein-Gordon equation

$$\Box\phi - RF'(\phi) + V'(\phi) = 0 \tag{162}$$

where $F'(\phi) = dF(\phi)/d\phi$ and $V'(\phi = dV(\phi)/d\phi$. Now it is important stressing that Eq.(162) is nothing else but the contracted Bianchi identity, so the above results hold demonstrating the consistency of the scheme. This feature shows that the effective stress-energy tensor at right hand side of (158) is a zero-divergence tensor and this fact is fully compatible with Einstein theory of gravity also if we started from a 5D-space. At the end, the reduction procedure preserves the features of standard General Relativity. Now we have the task of the physical identification of the fifth dimension. To this goal, let us recast the generalized Klein-Gordon equation (162) as

$$(\Box + m^2_{eff})\phi = 0, \tag{163}$$

where

$$m_{eff}^2 = [V'(\phi) - RF'(\phi)]\phi^{-1} \tag{164}$$

is the effective mass, i.e., a function of ϕ, where self-gravity contributions $RF'(\phi)$ and scalar field self-interactions $V'(\phi)$ are taken into account. In any quantum field theory formulated on curved space-times, these contributions, at one-loop level, have the "same weight" [84]. A natural way to generate the masses of particles can be achieved starting from a 5D representation and the concept of mass can be recovered as a geometric derivation. Finally, as we have seen, the reduction mechanism can select also $\epsilon = 1$ in the metric. In this case, the 5D-Klein Gordon equation (38), and the 5D-field equations (27) have wave-like solutions of the form

$$dS^2 = dt^2 - f(t, x_1)(dx^1)^2 - f(t, x_2)(dx^2)^2 - f(t, x_3)(dx^3)^2 + (dx^4)^2 \tag{165}$$

where

$$f(t, x_j) = exp\ i(\omega t + k_j x^j), \qquad j = 1, 2, 3. \tag{166}$$

It is very interesting to note that intrinsically such a solution has two times and, as a direct consequence, due to the structure of the functions $f(t, x_j)$, closed time-like paths are allowed. The existence of closed time-like paths means that Anti-De Sitter [96] and Goedel [97] solutions are natural possibilities in the OQR dynamics.

Taking again into account the previous arguments of par. 4.3 and 5.1 and substituting $m_{eff}^2 = m^2$, the splitting of Klein-Gordon equation developed (from Eq. (128 to Eq. (143)) gives rise to forward and backward evolution of (entangled) particles, i.e., it yields two time arrows together with the other features of particles. In fact, let us rewrite Eq. (142), which are also solutions of Eq. (54)

$$\psi(x) = e^{-ikx}u(k), \qquad \psi(x)^* = e^{ikx}u^*(k) \tag{167}$$

which can be interpreted, as we have seen, respectively, as progressive and regressive solutions. From the above calculations, it is straightforward that they contains all the information on the spin and the charge of the particle due to the fact that they derive directly from conjugate Dirac equations. A function (a superposition) of the form

$$\varphi(x) = \alpha_1\psi(x) + \alpha_2\psi^*(x) \tag{168}$$

where α_1, α_2, are constants, is a general solution of the 4D-dynamics and the states ψ and ψ^* can be interpreted as entangled since they can influence each other also when they are disconnected (see for the demonstration [86]). At the end, the absolute validity of conservation laws gives rise to a symmetric dynamics (backward and forward evolution of the system) and the entanglement of states is determined without arbitrary imposition [85, 90], in a completely general way and developed in the framework of any curved space-time. At this point, remembering the condition (52), we can recombine all the solutions which we have obtained through Eqs.(56) and (167). The field ϕ can be expanded as a Fourier series so that

$$\Phi(x^\alpha, x^4) = \sum_{n=-\infty}^{+\infty} \left[\psi_n(x^\alpha)e^{-ik_n x_4} + \psi_n^*(x^\alpha)e^{ik_n x_4}\right], \tag{169}$$

where it is important to stress that ψ and ψ^* are the 4D-solutions, while $e^{\pm ik_n x_4}$ are the embedded components coming from the reduction procedure. In general, every particle mass can be selected by solutions of type (56) while other particle features, as charge and spin, are selected by the other 4D solutions (167). It is worth to stress that the number $k_n x_4$, i.e., the ratio between the two lengths x_4/λ_n, fixes the interaction scale. Geometrically, such a scale is related to the curvature radius of the embedded 4D-space, where particles can be identified and, in this sense, our approach is an induced-matter theory, where the extra dimension cannot be simply classified as "compactified" since it yields all the 4D dynamics giving origin to the masses. Moreover, Eq. (169) is not a simple tower of mass states but a spectrum capable of explaining the hierarchy problem.

Finally also gravitational interaction scale can be discussed in this framework considering as fundamental the scale in which GN (Newton constant) is $G_N = \frac{^{(5)}G}{2\pi l}$, so then the Planck length

$$\lambda_p = l = \left(\frac{hG_n}{c^3}\right)^{1/2} \tag{170}$$

instead of the Compton length and this fixes the vacuum state of the system, since the masses of all particles can be considered negligible if compared with the Planck scales.

These results deserve a further comment due to the fact that the mass of a given particle is fixed by the embedding process, while the further splitting of 4D-Klein Gordon equation in two (forward and backward) Dirac equations, gives rise to the emergence of other features of particle as the spin and the charge, which are automatically conjugated and conserved by the occurrence of two time arrows. In terms of a group description, the splitting process defines the SU(2) (spin) and U(1) (charge) components, of the above considered supergroup $\mathcal{G}_5$.

5.4. Open Quantum Relativity Consequences on Time

The existence of closed time-like paths, means that Anti-De Sitter [96] and Goedel [97] solutions are naturally allowed possibilities in the dynamics of OQR and these results, in the context of time machine, have a deep meaning, since they are not "anomalies" as in the framework of the classical theory (General Relativity) but are "standard outputs" in the framework of this Open Quantum Relativity. While the concept of time travels in General Relativity is related to the relativistic effects among frameworks in relative motion on the same time arrow, here, in this approach, the relative motion can be taken into account on two time arrows, and this means that a closed time-like path does not need a topology change imposed "by hand", but can comes out by the projected motion along a circumference as combination of two linear motions in opposite sense, so then the topology change is a natural feature of Nature [93, 88], coming out every time a topology change is necessary to preserve a conservation law. It is clear that in OQR there is the full recovering of Special and General Relativity principles (because we do not need to travel faster than light) but the mechanism of relative motion between two frameworks, which is generating the hypothesis of time travels, is clearly not the same if applied on one or two time arrows [88]. The new scheme, provides first principles on which a Time machine should be based, because it is fixing the limits in which the phenomenon would be eventually possible. In fact, as we have seen, only in a situation of impossibility to avoid otherwise a violation of a conservation law, it happens that the Nature reacts changing the topology, so then "breaking" the Space-Time. The Entanglement is at the ground of this possibilit even at macroscopic level. In fact, there are works of authors [106] which demonstrate that an entanglement of two macroscopic systems does not need the entanglement of every component of them in correspondence one-to-one (which will be probably impossible to obtain) but only a correspondence at least under the limit of the uncertainty principle (evidently not easy, but perhaps not impossible).

In this framework, the Many Worlds Theory should be taken not as a possibility, but as a necessity, in order to avoid new and deeper paradoxes induced by backward time travels. In this picture, every possible universe can be reduced to a “local fact”. It is very difficult, today, to estimate all the consequences of the above considerations, mainly concerning a “time machine”, nevertheless one should try to derive some conclusions to open an useful debate. Time travel hypothesis in which two time arrows (and CTCs) are general features of Nature, represents a break point with the more traditional view of time machine, but able, in principle, to reconcile General Relativity and Causality in a framework in which Relativity and Quantum Mechanics are deeply connected. The Open Quantum Relativity comes out as a necessary evolution of a process which starts from the very basic principle and arrive to the generalization of entanglement concept as a gate, through topology changes, for motion in time [145]. The OQR results are that, in order to overcome problems of today physics, two dynamical solutions (forward and backward in time) are necessary and formally related to conservation principles.

6. Open Time

Now we are going to try to forecast the consequence of what we call an “Open Time”, but, before that, let us briefly recall what this concept means and where comes from. In OQR, paradoxes and shortcomings of contemporary physics, and the apparent dichotomy between Relativity and Quantum Mechanics can be settled, starting from the assumption that the conservation laws can never be violated. This principle leads, in fact, to the basic features that an unifying dynamical scheme can be formulated in a 5D space, and it exists a covariant symplectic structure for tensor, vector and spinor fields by which it is possible to construct suitable Hamiltonian scalar invariant. The reduction process of dynamics from 5D to 4D gives rise to several new features and some of them have observable effects. In 5D, it is possible to construct a description based on a stress-energy tensor where Bianchi identities hold in any case, since there are only kinetic terms without self-interacting potentials [91]. It is worth to note that a covariant symplectic structure can be found for every Hamiltonian invariant and curvature invariants of General Relativity can show such a feature. Another remark deserves the fact that, starting from such invariants, covariant and contravariant vector fields can be read as the configurations q^i and the momenta p_i of classical Hamiltonian dynamics, so then the Hamilton-like equations of motion are recovered from the application of covariant derivative to both these vector fields. Specifying the nature of vector fields, we select the particular theory, because the scheme is independent of the nature of vector field and, for us, this is a strong hint toward a unifying view of basic interactions, gravity included [93]. In this scheme, dynamics is singularity free and the reduction procedure to a 4D spacetime gives rise to a scalar-tensor theory of gravity where gravitational coupling is a function of a scalar field.

A very significant consequence of this general approach is that, in 4D, two time arrows and closed time-curves are naturally emerging, allowing a new interpretation of several phenomena, among them the entanglement [85] and the topology change is the mechanism assuring the conservation laws not only in EPR-type phenomena, but also in collapsing astrophysical systems (e.g. mass-energy for black holes [104]). This feature is fully compatible with Relativity since it can be dynamically realized in the 5D-4D reduction process considering a-luminal geodesics, and, moreover, it allows the full recovery of the essential concept of causality. The existence of two Time arrows (beside the other considerations), imply the existence, as a necessity, of a fifth dimension, to "open" both direction in Time and this is another radical change.

In conclusion, asking for a General Conservation Principle and a covariant symplectic structure, we have achieved a self-consistent theory with the minimal request of free parameters and dimensions: the Open Quantum Relativity. As a consequence of the theory, the role of time has to be revised toward a full recover of causality: this means that the absolute validity of conservation laws needs backward and forward causation and then two time arrows. This is the core of our scheme, which results in a unified view of modern science, no longer separated in quantum and classical physics and allows to frame in a comprehensive picture several shortcomings and paradoxes, really opening new perspectives.

6.1. Open Time Time Travel in the New Framework

The new OQR scheme provides first principles on which a Time machine should be based, because it is fixing the limits in which the phenomenon would be eventually possible. It is very difficult, today, to estimate precisely all the consequences of the above considerations, concerning a "time machine", nevertheless one should try to derive some conclusions to open a debate. Due to the fact that there are experimental evidences, not only for the well established standard effects (e.g. the lengthening of the life of accelerated elementary particles) but also for new ones (like the instantaneous teleportation via entanglement effects), it seems impossible simply close the doors to any time travelling hypothesis and so the necessity for a new theoretical ground, to frame this possibility, was unavoidable. OQR comes out as a straightforward evolution of a process which starts from real necessities. Namely : the solution of the EPR problem as a conflict between Relativity and Quantum Mechanics [67, 86]; the generalization of entanglement concept as a gate, through topology changes, to settle Time paradoxes [145]; and finally, the suggestion of a dynamical scheme for the unification of the different interactions [93]. In fact, from the first works on this topic [86, 88, 89], devoted to find the solutions to the conflict between the Locality Principle and Quantum Mechanics (in the light of Bell theorem and of experimental observations showing that Quantum Mechanics is basically correct), we found a new and general approach. The results that came out, after a long series of papers, are that, in order to overcome the paradoxes, two solutions (forward and backward in time) are necessary and deeply related to conservation principles given by the Bianchi identities.

If this is the background, let us now remember the four classes of possibilities taken into account in this work on time and time travels. 1) Travels in time forbidden by a conjecture which seems reasonable, because based on the common sense which asks for the reason of the apparent lack of time voyagers, but also a conjecture not mathematically founded (it comes out from semi-classical approximations which are inadequate in a quantum world) and closer to a philosophical request than to a physical exhaustive demonstration; 2) time travels exotically based on particles, the tachyons, very difficult to define without violating Special and General Relativity and epistemologically unsatisfactory, if considered as not observable by definition; 3) time travels confined in the rigorous limit of General Relativity with, up to now, a great difficulty to solve the shortcomings questioning their existence; 4) finally, time travels coming out from a new extended theory of gravity, the OQR theory, in which two time arrows (and CTCs) are general features of the Nature, theory which represents a break with the traditional points of view, but able in principle to reconcile General Relativity and Causality in a framework where Relativity and Quantum Mechanics are deeply connected. These are, in summary, the four possible issues on time travel topic, which is questioning the scientific community, because it involves the whole problem of time definition. In any case, we do not believe that any semi-classical approximation will be enough to settle this modern puzzle. On this topic, any scientific consideration has to be seriously taken into account, but avoiding that the search to solve the existing paradoxes results in introducing theories involving too many free parameters and dimensions (eventually introducing even more paradoxes and shortcomings). At the end, we saw the necessity of a Quantum Relativity as an issue which was impossible to postpone any longer. A General Conservation Principle, together with two time arrows, the generalized entanglement concept in a Many Worlds framework; a fifth dimension to "open" the Relativity and a set of equations capable of giving dynamics of different phenomena like teleportation, black holes, gamma ray bursts and many others; are all parts of an approach which seems to us necessary to understand what happens in modern physics. Born to settle the contradictions in contemporary physics, at the end the OQR opened the way to a new discussion on Time Travels.

6.2. Open Time Many Worlds and Spacetiime Tunnels

In this framework, the Many Worlds Theory should be taken not as a possibility, but as a necessity, in order to avoid new and deeper paradoxes induced by backward time travels, like the famous hypothesis of a time traveller killing his grandmother when she was young (before his own father birth) so deleting himself from the origin and making impossible to kill grandmother. Only the Many Worlds Theory, in its full version [1] of an infinity of universes (each one representing a virtual possibility of evolution which can become a reality), can settle this puzzle, saving time travel generalized hypothesis, causality and pure logic together. In this picture, a "physical" universe (i.e., from the Big Bang up to where it is possible conceive to get at light speed) can be reduced to a "local fact" and the "whole" Universe it would be the sum of all the infinite universes corresponding to every virtual possibility, a sort of Universe of Universes. More properly, the Quantum Cosmology is the quantization of dynamical systems which are "universes" and, in this context, the Universe, supposed as a whole (the ensemble of all the possible universes), has a quantum mechanical nature and so an observable universe is a limit concept valid in particular regions of a manifold (superspace) composed by all the possible space-like 3-geometries and local configurations of the matter fields. The purpose of Quantum Cosmology is to relate all the measurable quantities of the observable universe [1] to the assigned boundary conditions for a wave function in the superspace. We must

[1]An operative definition of "observable universe" could be an universe where cosmological parameters as the Hubble one H_0, the deceleration parameter q_0 , the density parameters Ω_M, Ω_Λ, Ω_k and the age t_0 can be inferred by observations [112].

stress that, a singularity free theory implying the general conservation principle, as the OQR, is a necessity to make of the Many World an "observable theory", as we are going to see. Specifically, conservation laws in phase space of physical systems (e. g. minisuperspace) give rise to natural selection rules by which it is possible to discriminate among physical and unphysical solutions which, in the case of Quantum Cosmology, can be seen as physical and unphysical universes. In a previous paper [1], we have shown that the reduction procedure of dynamics can give rise to a splitting of the phase-space of a physical system, by which it is possible to achieve the complete solution of dynamics. Leaving to the quoted paper for the formal derivation, this result can be applied to Quantum Cosmology, leading to the outcome that physical many worlds can be related to integrable multi-spaces of the above splitting. As a remark, the general solution (151) can be interpreted as a superposition of particular solutions (the components in different directions) which result solved (i.e., separated in every direction of configuration space) if a related symmetry exist. The role of symmetries and conservation laws is prominent to interpret the information contained in the wave function of the universe, which is solution of the WDW equation [108]; in fact, the conserved momenta, related to some of the physical variables defining the minisuperspace, select oscillatory behaviors (i.e., strong peaks) in Ψ (wave function, semiclassically defined) which means correlations among the physical variables and then classical trajectories, whose interpretation is that of "observable universes". The Hartle criterion gives an interpretation of such correlations and, in particular, if the wave function is strongly peaked in some region of configuration space, the correlations which characterize such a region can be observed. If strongly peaked (oscillatory) subsets of the wave function of the universe are found, then one (or more) conservation law is present for a given minisuperspace model, so oscillatory parts of the wave function can be always connected to conserved momenta and then to symmetries. The Hartle criterion works in the context of an Everett-type interpretation (e.g. Many Worlds interpretation) of Quantum Cosmology, which assumes the idea that the universe branches into a large number of copies of itself whenever a measurement is made. In other words, the main issue of quantum cosmology is to search for boundary conditions in agreement with the astronomical observations and these conditions have to be contained in the wave function of the universe $|\Psi\rangle$ (defined now as a quantum state). The dynamical behavior of $|\Psi\rangle$ in the superspace is described by he Wheeler-DeWitt (WDW) equation [108] and the Hartle criterion [61] is the interpretative scheme for the solutions of the

WDW equation, otherwise difficult to handle because with an infinite degree of freedom. At the end, if conservation laws are not valid the WDW multi-space solution give rise to non-observable universes (technically known as Instanton solutions). Furthermore, the existence of conserved quantities always implies a reduction of dynamics, which means that the order of equations of motion is reduced thanks to the existence of first integrals. In fact, phenomena appearing in standard physics as due to symmetry breakings can be encompassed in a multi-space formulation [1]. In other words, the need of a multi-space formulation of the theory, gives rise to a direct application of the "Many Worlds" Interpretation of Quantum Mechanics, in the sense that multispaces are nothing else but many worlds in the framework of Quantum Cosmology [1]. The configuration spaces derived from the request of integrability of the dynamical systems are physical universes, (i.e., observable universes) where cosmological parameters can be observed. So then, if conservation laws are not present, universes which come out in a Many Worlds interpretation are "unphysical" that is, it is not possible to label them by a set of observable cosmological parameters. Before to go to the general conclusions, let us sketch very briefly and from a purely qualitative point of view (the only one possible at this stage) some simple hint on the possible use of a Natural Time tunnel or on the design of a concevaible human made Time Machine (or Entanglement Machine). Even if it is apparently the easiest to conceive, the idea of the use of a possible "Natural" Time Machine, like the space-time tunnel of the system Black Hole - Warm Hole - White Hole, it is very probably extremely difficult to realize and manage (perhaps worst than the artificial ones) because of the enormous gravitational attraction in the proximity of the B.H. that can destroy every structure approaching it (whatever type it is) much before of the beginning of tunnel, without a spacecraft able to cross the final distance in $\Delta t = 0$, which is very difficult to conceive. Also for an artificial mechanism, the starting point is always an otherwise unavoideble violation of a conservation law, to force the nature to react changing the topology (so breaking the space-time). Now, without obviously any claim of completness or formal treatment, an attempt to conceive such a violation, on the energy, con be imagined trying to think, for instance, to an ideal transformer and utilize (or invent) a mechanism able to furnish energy so quickly that the secondary cannot dissipate it, even trough an explosion. A purely conceptual device would be, as a naive (and perhaps impossible) example, to use an instantaneus (in the limits of uncertainty principle) transition of phase using an big amount of energy and reducing the time enough to reach the asked for conditions. Is it possible, today, for the physics principles and the existent technologies, transfer energy so instantaneously to avoid any possibility to dissipate it, so realizing the unavoidable violation of a conservation law necessary for an entanglement machine? We will see. At this stage, we must accept the fact that, even to simply conceive a pure Gedanken Time Machine, the road is very long, but, conversely, the principles are very clear. Whatever type of energy and whatevere type of transferring mechanism one can suggest, the point is always the same: create a situation of violation of a conservation law, unavoidable with the mechanism of the standard physics. Another conjecture can also be envisaged on conservation of quantum number. These are even not schemes, but just hints on the road toward a "gedanken Entanglement Machine" and this road needs competences in practically all the branches of physics and Science, so it must be considered also a "call for ideas". But, if we are still far from the design of a Time Machine, the principles, nevertheles, seems fixed.

6.3. Open Time Up to the Last Consequences

After the presentation of the main features of this new approach, and reminding to the quoted papers for the full treatment, let us now stress here the results on the specific topic of this paper: how OQR works for the time and time machine issues. It is clear that once accepted the idea of the natural existence of a second time arrow backward oriented (beside the CTCs), the full definition of a time machine is changing, because, in this case, it will be not a exception coming out from very particular conditions, but a possibility coming from general feature of the Nature and the problem is now what to do to obtain it. In fact while the concept of time travels in General Relativity is only derivable from different frameworks on the same Time arrow, here, in the new approach, the relative difference can be directly derived from two Time arrows and this means that a closed timelike path does not need to exist, a topology change imposed by hand (like the change from longitudinal to circular time coordinate in Goedel cylindrical representation), because the projected motion along a circumference is the combination of two linear motions in opposite sense, so then the topology change possibility, in OQR is the natural result of an underlying feature of Nature. This feature is really providing a dynamics of what Schröedinger called Verschränkung (translated by himself as entanglement in a letter to A. Einstein). It is evident that there is the full recovery of the Relativity (because we do not need to travel faster than light) but here we are in the frame of an

"Open" Quantum Relativity. The new theory, even not entering in suggestions for the technical devices of an hypothetical time machine, provided nevertheless first principles on which such a machine should be based, because it is fixing the limits in which such phenomenon will be eventually possible. Due to the fact that only in a situation of impossibility to avoid a violation of a conservation law, it happens that the nature reacts changing the topology, we should first of all find in the universe such a situation or be able to recreate it. As we have seen, in the first case (in the today knowledge), the laboratory for such an experiment should be a black hole, in the second case, an entanglement machine. This would be a kind of machine very difficult to design technically yet, but that we cannot in principle exclude. In fact, as we know, there are works of authors [106] which demonstrates that an entanglement of two macroscopic systems does not need the entanglement of every component of them in correspondence one-to-one but only a correspondence at least under the limit of the uncertainty principle. In both cases, natural or artificial, a quantitative method of calculation of the arrival spacetime coordinates is then necessary, on the basis of the exposed theory. Under the realistic hypothesis that the traveller wants to come back (and stay alive), two further conditions are then necessary: i) the hypothetical spacetime craft must move together with the voyager, because the existence of two time arrows can make a time travel independent, without an automatic back ticket and moreover, ii) the approach to a time tube should be not simply as fast as possible (and always relativistic) but instantaneous, (a-luminal) to avoid that the travellers could be squeezed in elementary particles (one can postulate a null pressure in a wormhole, but, before getting it, in the black hole proximity the gravity is destroying the structure of any conventional object).

It is very difficult, today, to estimate all the consequences of the above considerations concerning a Time Machine, nevertheless one should derive some conclusions to open an useful discussion. Nobody can say, now, when a time machine will be realized in the future, but, in any case, it has to obey, at least, to the following criteria from the above principles:

1. It has to be possible to reach critical conditions of violation of a conservation law, in such an otherwise unavoidable way to give rise (to evade it) to the topology change phenomenon, which is at the basis of spacetime tunnelling possibility.

2. It has to be possible the existence of spacetime regions which can be

considered entangled as a whole, in the sense that the fraction of their components, not mutually entangled one-to-one, has to be lower of the limit of uncertainty principle.

3. It has to be possible that what is going to be involved in a spacetime tunnelling (from a simple information up to a large spacetime craft) can be considered only a perturbation of the entangled regions.

4. It has to be possible to be in the framework of Many Worlds Theory (i.e., this theory has to be true), because even the simple fact of traveling back in time violates the causality law of our Universe, the recovery of which being assured by the virtual existence of several universes as many as the possibilities of changing.

5. It has to be possible to estimate the probability of permanence of entanglement conditions between two worlds "tunneled together", i.e., to estimate when the cascade effects of initial perturbations change enough the evolution of the arrival world to exceed the uncertainty limit, so disconnecting the two spacetime regions.

6. It has to be possible to calculate exactly the spacetime region where and when the travel is going to end (i.e., well posed Cauchy boundary conditions).

7. It has to be possible to know how realize the inverse operation to come back home. In any case, it seems necessary that the technical device (whatever it will be) that initially produces the requested conditions for the topology change, should travel with the passenger, because the inverse operation is not generally automat but should obey to the same laws.

8. It has to be possible a non-destructive topology change inducing mechanism (i.e., not undergoing through a process of separation and recombination, eventually incompatible with life) but a mechanism able instead to apply the transformation on the system as a whole.

9. It has to be possible a topology change mechanism almost instantaneously starting up (in the limit of Heisemberg uncertainty principle for energy and time) to get the time-tunnelling.

10. It has be possible technically conceive a gedanken spacetime craft (and then a real one) in order to use natural time-tunnels or a gedanken time machine to realize artificial time-tunnel.

These are the necessary condition to "travel in Time" in our theory, but what really means, how can be represented this possibility in our picture? In fact one of the major difficulties to really understand the meaning of the expression travel in Time, is due to the fact that the "visualisation" of what would happens is difficult for people sensorially, historically and philosophically prepared to a Time flowing always in the same orientatione toward what we call future.

In order to try to overcome that, with a naive but perhaps useful example, let us think in terms of a classical film, made with a succession of images on a plastic support. Now it is evident that if we run the film in the contrary sens, we obtain that the "future", in the projection of the movie story, is in fact the Past. Which means, if you want, that the "local" time of the film, can be independent of the "global" time of the observer. The essential point, of this situation, is based on a very important characteristic: the persistence of the images on the plastic support. So the "Past", in the close system of a movie film, is not destroyed. In some sense it is what happens in Open Quantum Relativity (OQR) where the CTC (Closed Time-like Curve [16, 42]) allow to repeat the past, because Time is going again on the same foot. In this picture, the operation to play the film in the opposite sense, is the equivalent of a change of topology with related inversion of Time Arrow. Quite interesting, in this picture, it is also to compare the fact that, to see a movie, it is necessary that the capability of human eye to fix on time images should be less than the speed with which the images themselves run, because, if not , we would see a series of photographies instead of a movie, and this, in some very relative sense, is what happen with human inability to experience this double Time arrow. Finally it is usual to play a film in the normal sense (i.e., from the part considered as the past, to the part which is the future of the film story) and if we stop and go back, normally we do that with a faster and drastic operation (without seeing the film) than we restart in the usual direction in normal speed. It is a pictorial point of view in some sense representing the operation to going back to the past, considered as an instantaneous non linear operation (happening when the Nature is forced to change the Topology) followed by a "normal" linear evolution from the past to the future, but starting from the new initial point. Anyway the crucial point is the "persistence" of the Past, which is not disappearing (like a photogram in a movie) how we are instead traditionally used to think, but it persists and can be changed (creating a new parallel universe) like if we cut or add photograms in the film creating a new one, before to see it again. This is of course only an attempt to help to imagin what we are going to expect in the frame of OQR (an attempt at the end necessarily inaccurate and largely incorrect) but an useful attempt, because it is so difficult to conceive an idea which is so far of our human experience, that we estimate that even a rough example, was good to introduce the discussion. But at the end only the definitions, the dinamical physical laws and the mathematical formalism are the real ground of the new approach, which, developped by us for more than 15 years, bring to a new theory at the intersection point between General Relativity and Quantum Mechanics, connecting intrinsecally them and and focusing on Time Travels hypothesis.

7. Conclusion and Perspectives

The concept of a "variable time" is really the most revolutionary concept of the entire history of science, because it provided a completely new point of view for physics and opened the doors to all the following speculations on the nature of time, until the extreme hypotheses of time travels. Now, a theory (not a pure conjecture) that allows time travels can change not only our knowledge of the Universe, but also and dramatically the way in which we see ourselves in such an Universe and, at the end, the whole meaning of human life and that is the reason why, all along the last Century, so many works have been devoted to this subject. But any attempt in this direction was unsuccessful mainly because of a very fundamental problem: the absence of a general theory capable of unifying General Relativity and Quantum Mechanics, i.e., a real Quantum Relativity. The Open Quantum Relativity is what we propose as answer to the search for this general theory. In an overview of modern physics, several shortcomings and paradoxes, and this dichotomy between relativity and quantum mechanics, are framed, and eventually solved, assuming as General Conservation Principle the fact that conservation laws can never be violated. This principle leads to some basic features: i) a general unifying dynamical scheme can be formulated in a 5D space seen as the lowest dimensional necessary extension; ii) a symplectic covariant structure can be sought for every vector or tensor field by which it is possible to construct a scalar invariant; iii) two time arrows emerge as direct consequence of dynamics. In 5D, it is possible to construct a description based on a stress-energy tensor for a generalized scalar field where Bianchi identities must hold in any case and, in this scheme, dynamics is singularity free and the reduction procedure to a 4D-spacetime gives rise to a scalar-tensor theory of gravity, where gravitational coupling is a function of a scalar field.

A very significant consequence of this general approach is that, in 4D, two time arrows and closed timelike curves are exact solutions of dynamics allowing a new interpretation of several phenomena, among them the entanglement. In fact, entangled systems are not hidden characteristics of Quantum Mechanics but superpositions of progressive and regressive solutions in a dynamical framework were backward and forward causation are booth existent, without any violation of conservation laws. In this sense, topology change is the mechanism capable of explaining the entanglement, and this feature is fully compatible with relativity, since it can be dynamically realized, in the 5D-4D reduction process, considering a-luminal geodesics. Finally, we have to stress that a covariant, symplectic structure can be found for every Hamiltonian invariant. In fact, following the same scheme of Quantum Mechanics, any theory of physics has to be endowed with a symplectic structure in order to be formulated at a fundamental level, so the scheme is very general. We pointed out that curvature invariants of general relativity can show such a feature and, furthermore, they can be recovered from Hamiltonian invariants opportunely defined. What is crucial, in our approach, is that it comes out an unified view of modern physics (no longer sharply separated in "quantum" and "classical") allowing to frame in a comprehensive picture the law of the physics, so opening the perspective of a new lecture of the Nature.

The general validity of OQR, ranging from particles physics to astrophysics and cosmology, and its capability to reproduce correctly the experimental data wherever it has been applied until now, including the new phenomena (which cannot be a coincidence) is one of the more significant characteristic of this theory. But it is in Time definition that the new perspective, due to the Open Quantum Relativity, can show the more fundamental consequences. The two Time arrows, really can open the door to new and dramatic changements and perhaps, in the next future, the word Time Machine will no longer indicate only a "Closed Timelike Curve", but a much more general and evocative reality. This is certainly our hope, but at the same time is also a formal, straightforward, result. At the end of this long excursus (and of the long series of works dedicated by us all along the years) we think that we can not avoid the duty to answer clearly to the question, which is naturally emerging: is a Time Travel really concevaible? Yes it is.

References

[1] G. Basini, S. Capozziello, *Progress in Physics*, V. 4, 65, (2006).
F. Smarandache. *A unifying field in logics*. Am. Res. Press, Rehoboth, (2003).
J.B. Piran T., Weinberg S., World Scientific, Singapore, (1991).
J.J. Halliwell, *Nucl. Phys. B*266, 228 (1986).
J.J. Halliwell, in *Quantum Cosmology and Baby Universes*, Eds.
S. Coleman S., Hartle H. Everett, *Rev. Mod. Phys.* Vol. 29, 454, (1957).

[2] M. Green, J. Schwarz, E. Witten, *Superstring Theory*, Univ. Press, Cambridge 1987.

[3] S. Ferrara and P. van Nieuwenhuizen, *Phys. Lett. B*78, 573 (1978).
S. Ferrara and P. Fayet, *Phys. Rep. V.* 32C, 250 (1977).
P. van Nieuwenhuizen, *Phys. Rep.* 68, 189 (1981).

[4] S.W. Hawking, *Phys. Rev. D* 462 (1992).

[5] M.Visser, in The Future of theoretical physics and cosmology, *Cambridge Proceedings*, Cambridge Univ. Press, Cambridge (2002).

[6] S.V. Krasnikov, *Phys. Rev.* D54, 7322 (1996).

[7] J.R. Gott, *Phys. Rev. Lett.* 661126 (1991).

[8] M.S. Morris, K.S. Thorne, and U. Yurtsever, *Phys. Rev. Lett.* 61, 1446 (1988).

[9] S.Deser, R.Jackiw, and G. Hooft, *Phys. Rev. Lett.* 68, 267 (1992).
S.Deser, R. Jackiw, *Comments Nucl. Part. Phys.* 20, 337 (1992).

[10] W. Huang, *Phys. Rev.* D60, 067505 (1999).

[11] D.G. Boulware, *Phys. Rev.* D464421 (1992).

[12] T. Tanaka et al., *Phys. Rev.* D524503 (1995).

[13] M. Lyutikov, *Phys. Rev.* D494041 (1994).

[14] M.J. Cassidy, *Class. Quantum Grav.* 121685 (1995).

[15] P. Gonzalez-Diaz, *Phys. Rev.* D566293 (1997).

[16] Li Xin Li Class. *Quantum Grav.* 132563 (1996).

[17] J.J. Sakurai, *Modern Quantum Mechanics*, Addison-Wesley Publ. Co., N Y, 1994.

[18] J. Bell, *Rev. Modern Phys.* 38, 447 (1966).
J. Bell, *Physics* 1, 195 (1965).

[19] A. Aspect, P. Grangier, G. Roger, *Phys. Rev. Lett.* 47, 460 (1981).
A. Aspect, P. Grangier, G. Roger, *Phys. Rev. Lett.* 49, 91(1982).
A. Aspect, J. Dalibard, G. Roger, *Phys. Rev. Lett.* 49, 1804 (1982).

[20] C.H. Bennett et al. *Phys. Rev. Lett.* 70, 1985 (1993).
M. Zukowski et al., *Phys. Rev. Lett.* 71, 4287 (1993).
H. Weinfurter, *Europhys. Lett.* 25, 559 (1994).

[21] D. Bouwmeester et al. *Nature* 390, 575 (1997).

[22] C.H. Bennett, *Phys. Rev. Lett.* 70, 1895 (1993).

[23] A.Shymony, Foundation of Quantum Mechanics in the light of new technology, Kamefuchi et al. Eds, *Phys. Soc. of Japan*, (1984).

[24] L. Randall and R. Sundrum, *Phys. Rev. Lett.* 83, 4690 (1999).

[25] E. Recami, La Riv. *N. Cim.* 9, No. 6 (1986).

[26] R. Tomaschitz, *Int. Journ. Mod. Phys. D*, Vol. 7, No. 2, 279 (1998).

[27] C.W. Misner and J. A. Wheeler, *Ann. Phys.* 2, 525 (1957)

[28] J.A. Wheeler, *Geometrodynamics*, Academic Press, New York (1962).

[29] A. Einstein and N. Rosen, *Phys. Rev.* 48, 73(1935).

[30] D. Gannon, *J. Math. Phys.* 16, 2364 (1975).

[31] C.H. Lee, *J. Math. Phys.* 17, 1226 (1976).

[32] G.J. Galloway, *J. Phys.* A16, 1435 (1983).

[33] J.L. Friedman, K. Schleich, and D.M. Witt, *Phys. Rev. Lett.* 71, 1486 (1993).

[34] K. S. Thorne, in GRG13: *Proceedings of the 13th Int. Conf. on General Relativity and Gravitation*, Cordoba, Argentina, 1992, Bristol Inst. of Physics, p.295.

[35] M. Visser Lorentzian Wormholes:From Einstein to Hawking, A. I. P. Press (1995).

[36] E. Flanagan and R.M. Wald, *Phys. Rev.* D54, 6233 (1996).

[37] I. D. Novikov, *JEPT* 95, 769 (1989).

[38] J.L. Friedman et al. *Phys. Rev.* D42, 1915 (1990).

[39] V.P. Frolov and I.D. Novikov, *Phys. Rev.* D42, 1057 (1990).

[40] R. Geroch, *J. Math. Phys.* 12, 918 (1971).

[41] W.J. van Stockum, *Proc. Roy. Soc. Edin.* 57, 135 (1937).

[42] K. Goedel, *Rev. Mod. Phys.* 21, 447 (1949).

[43] F. de Felice, *Nuovo Cim.* 65, 224 (1981).

[44] G. Klinkhammer, *Phys. Rev.* D46, 3388 (1992).

[45] A. Folacci, *Phys. Rev.* D46, 2726 (1992).

[46] R.M. Wald and U. Yurtsever, *Phys. Rev.* D44, 403 (1991).

[47] L.H. Ford and T.A. Roman, *Phys. Rev.* D41, 3662 (1990).

[48] R. Geroch, *J. Math. Phys.* 8, 782 (1967).

[49] I. D. Novikov, *Evolution of the Universe*, Cambridge Univ. Press, p.169 (1983).

[50] Ya.B. Zel'dovich and I.D. Novikov, Relativistic Astrophysics, 2: *The Structure and the Evolution of the Universe*, Univ. of Chicago Press, Chicago (1975).

[51] I.D. Novikov, *Phys. Rev.* D45, 1989 (1992).

[52] F. Echeverria, G. Klinkhammer, and K.S. Thorne, *Phys. Rev.* D44, 1077 (1991).

[53] E.V. Mikheeva and I.D. Novikov, *Phys. Rev. D* 47, 1432 (1993).

[54] A. Lossev and I.D. Novikov, *Class. Quantum Grav.* 9, 2309 (1992).

[55] A. Carlini et al. *Int. Journ. Mod. Phys. D* 4, 557 (1995).

[56] A. Carlini and I.D. Novikov, *Int. Journ. Mod. Phys.* D5, 445 (1996).

[57] S.W. Kim and K.S. Thorne, *Phys. Rev.* D43, 3929 (1991).

[58] V.P. Frolov, *Phys. Rev.* D43, 3878 (1991).

[59] H.D. Politzer, *Phys. Rev.* D46, 4470 (1992).

[60] J.L. Friedman, N. Papastamatiou, and J.Z. Simon, *Phys. Rev.* D46, 4442 (1992).

[61] J.B. Hartle, *Phys. Rev.* D49, 6543 (1994).

[62] D. Deutsch, *Phys. Rev.* D44, 3197 (1992).

[63] D.S. Goldwirth et al. *Phys. Rev.* D49, 3951 (1994).

[64] P. Kanty, B. Kleihaus, *Phys. Rev. L.* 107, (2011);
I. Novikov, A. Shatskiy, *JETP* 114 (2012);
K. Bronnikov, L. Lipatova, A. Shatskiy. arXiv: 1312.6929v1 [gr-qc]25 D. (2013).

[65] G. Basini, S. Capozziello, *Mod. Phys. Lett. A* 20, 251 (2005).

[66] G. Basini, S. Capozziello, *Int. Journ. Mod. Phys. D* 15, 583 (2006).

[67] A. Einstein, B. Podolsky, N. Rosen, *Phys. Rev.* 47 (1935) 777.

[68] S.O. Alexeyev, M.V. Sazhin, *Gen. Relativ. Gravit.* 30 (1998),1187.

[69] S. Alexeyev, A. Barrau, G. Boudoul, M. Sazhin, *Cl. Q. Gravity*, 19 (2002) 4431.

[70] J.D. Bekenstein, Black Hole Thermodynamics, *Physics Today*, 1980.

[71] C. Quigg, *Gauge Theories of Strong, Weak, and Electromagnetic Interactions*, Addison-Wesley, Reading, MA, 1983;
P. Collins, A. Martin, E. Squires, *Particle Phys. and Cosmology*, Wiley, N. Y., 1991.

[72] G. Basini, P. Spillantini, Venezia, *Phys. in Space Station*, 1988, Ruffini Ed. SIF pr.17; G. Basini, A. Morselli, M. Ricci, La Riv. del N., *Cimento* 12 (1989) 4.
G. Basini, M.T. Brunetti, A. Codino et al. *Astrophysical Journal*, 467, L33(1996) G. Basini, S. Capozziello, G. Longo, *Rivista Nuovo Cimento*, 25, 11, (2002).

[73] A. Zichichi, La Riv. del N., *Cimento* 24 (2001) 12.

[74] A.D. Dolgov, M.V. Sazhin, Ya.B. Zeldovich, *Basic of Modern Cosmology*, Editions Frontieres, Gif-sur-Yvette, 1990.

[75] G. Basini and S. Capozziello, in: *Classical and Quantum Gravity Research*, Nova Science Publishers, Ed. M. Christiansen and T. Rasmussen, pp. 1-75, (2008).

[76] E.W.Kolb, M.S. Turner, *The Early Universe*, Addison-Wesley, Redwood, 1990.

[77] C.H. Bennett et al. *Phys. Rev. Lett.* 70, 1985 (1993); M. Zukowski et al., *Phys. Rev. Lett.* 71, 4287 (1993); H. Weinfurter, *Europhys. Lett.* 25, 559 (1994).

[78] D. Bouwmeester et al. *Nature* 390, 575 (1997).

[79] F. Selleri, Conc. *Foundations of Quantum Mechanics*, Menlo Park, CA, 1988;
F. Selleri (Ed.), *Quantum Mechanics versus Local Realism*, Plenum, London, 1988;
J. Bell, *Rev. Mod. Phys.* 38 (1966) 447;
J. Bell, *Physics* 1 (1965) 195.

[80] D. Bohm, *Quantum Mechanics*, Prentice-Hall, Englewood, N.J. (1951).

[81] A. Aspect, P.Grangier, G.Roger, *Ph. Rev. Lett.* 47,460(1981);
A. Aspect, P. Grangier, G. Roger, *Phys. Rev. Lett.* 49, 91 (1982);
A. Aspect, J. Dalibard, *Phys. Rev. Lett.* 49, 1804 (1982).

[82] C. Itzykson, J.B. Zuber, *Quantum Field Theory*, McGraw-Hill, Singapore, 1980.

[83] M. Kaku, *Quantum Field Theory*, Oxford Univ. Press, Oxford, 1993.

[84] N.Birrell, P.C.Davies, *Quantum Fields in Curved Space*, Cambridge Univ.1984.

[85] G. Basini S. Capozziello, *Europhys. Lett.* 63, 166 (2003).

[86] G. Basini, S. Capozziello, G. Longo, *Gen. Relativ. Grav.* 35 (2003) 189.

[87] L.D. Landau, E.M. Lifshitz, *Theorie du Champs*, Mir, Moscow, 1960.

[88] G. Basini, S. Capozziello, La Riv. del N., *Cimento* 27 N. 11 (2004).

[89] G. Basini S. Capozziello, *Europhys. Lett.* 63, 635 (2003).

[90] G. Basini, S. Capozziello, G. Longo *Phys. Lett.* 311 A, 465 (2003).

[91] G. Basini and S. Capozziello, *Prog. in Phys.* 3, 36 (2007).

[92] S. Weinberg, *Gravitation and Cosmology*, Wiley, New York (1972).

[93] G. Basini, S. Capozziello, *Gen. Relativ. Grav.* 35, 2217 (2003).

[94] D. Youm, *Phys. Rev. D* 62, 084002 (2000).

[95] J.E. Campbell *A Course of Differential Geometry*, Clarendon, Oxford (1926).

[96] D.H. Satinger, O.L. Weaver, *Lie Groups and Algebras with Applications to Physics, Geometry and Mechanics*, Springer-Verlag, Berlin (1986).

[97] K. Goedel, *Rev. Mod. Phys.*, 21, 447 (1949).

[98] C. Brukner, M. Zukowski, A. Zeilinger, quantum-ph/0106119 (2001).

[99] M. Visser, Lorentzian Wormholes: From Einstein to Hawking, *Am. Inst. Ph.* (1995).

[100] J.L. Friedman et al., *Phys. Rev. D* 42, 1915 (1990).

[101] S.W. Hawking, *Phys. Rev. D* 46 2 (1992).

[102] V.P. Frolov and I.D. Novikov, *Phys. Rev. D* 42, 1057 (1990).

[103] F. de Felice, Il N., *Cimento* 65, 224 (1981).

[104] G. Basini, S. Capozziello, G. Longo, *Astrop. Phys.* 20, 457 (2004).

[105] G.Basini, S. Capozziello, M. Ricci, F. Bongiorno, *Int. Journ. Mod. Phys.* D13, (2004);
S. Capozziello, M. Francaviglia, *Gen. Rel. Grav.* 40, 357, 420 (2008);
S. Capozziello, M.F. De Laurentis, *Phys. Rept.* 509, 1,154 (2011);
S. Nojiri, S. Odintsov. *Phys. Rept.* 505, 59-144, (2011).

[106] B. Julsgaard, A. Kozhekin, and E.S. Polzik, *Nature* 413, 400 (2001).

[107] A. Palatini, *Rend. Circ. Mat. Palermo* 43(1919), 203.

[108] B.S. DeWitt, *Phys. Rev.* 160 (1967) 1113;
C. Misner in *Relativity* eds. Carmeli, Fickler and Witten (Plenum Pub. Co1970);
C. Misner in *Magic Without Magic* ed. J. Klauder, W.Freeman, San Franc. (1972).

[109] A. Ashtekar, *Phys. Rev. Lett.* 57 (1986) 2244;
A. Ashtekar, *Phys. Rev. D* 36 (1987) 1587;
A. Ashtekar Banff, *Workshop on Grav. Phys.* (1990) and references therein.

[110] J.B. Hartle in Gravitation in Astrophysics Cargese, 1986, eds.
B. Carter and J. Hartle, Plenum, New York, (1986);
S.W. Hawking and D. *Page Nucl. Phys. B* 264 (1986) 185.

[111] G.Basini and S.Capozziello, *Journal of Modern Physics* 1, (2010);
J.B. Hartle, Proc.of 23rd Solvay Conference, *The Quantum Structure of Space and Time*, Eds. D. Gross, M. Henneaux, World Scientific, Singapore (2007).

[112] G. Basini, S. Capozziello, F. Bongiorno, *Int. Jou. Mod. Phys.* D 13, 717 (2004).

[113] S.M. Caroll, W.H. Press, E.L. Turner, *Ann. Rev. Astr. Ap.* 30, 499 (1992).

[114] G. Basini, S. Capozziello, *Astrop. Phys.*, 21, 543 (2004).

[115] B.P. Schmidt et al. *Ap. J.* 507, 46 (1998);
A.G. Riess et al., *Ap. J.* 116, 1009 (1998);
S. Perlmutter et al. *Ap. J.* 483, 565 (1997);
S. Perlmutter et al. *Nature* 391, 51 (1998).

[116] P. de Bernardis e tal. *Nature* 404, 955(2000);
R. Stompor et al., *Ap. J.* 561,L7(2001);
D.N. Spergel et al. astro-ph/0302209 (2003).

[117] M. Tegmark et al. astro-ph/0310723 (2003).

[118] Planck Collaboration, *Astron. Astrophys.* 571, A22 (2014).

[119] Y. Wang, *Ap. J.* 536,531 (2000).

[120] M. Birkinshaw, *Phys. Rep.* 310(1999) 97.

[121] M. Schmidt, *Ap. J.* 535, 117 (1999).

[122] J.I. Katz and L.M. Canel, *Ap. J.* 471, 915 (1996).

[123] S. Covino et al, *Astron. Astrophys.* 348, L1 (1999).

[124] R.A.M.J. Wijers et al, *Ap. J.* 523, 33L (1999).

[125] J. van Paradijs et al, *Nature*, 386, 686 (1997).

[126] *T. Piran Phys. Rep.*, 314, 575 (1999); *T. Piran Phys. Rep.*, 333, 529 (2000).

[127] G. Ghisellini, D. Lazzati, Mon. Not. R. Ast. Soc. 309, L7 (1999).

[128] M.J. Rees, P. Meszaros, *Ap. J.* 430, L93 (1994).

[129] R. Sari, *Ap. J.* 524, L43 (1999).

[130] K. Stelle, *Gen. Rel. Grav.* 9 (1978) 353.

[131] R.H. Sanders, *Ann. Rev. Astr. Ap.* 2 (1990) 1.

[132] M. Kenmoku, Y. Okamoto, and K. Shigemoto, *Phys. Rev.* 48 D (1993) 578.

[133] J.D. Anderson et al. *Phys. Rev. Lett.* 81 (1998) 2858.

[134] I. Quant and H.-J. Schmidt, *Astron. Nachr.* 312 (1991) 97.

[135] C. Will, *Theory & Experiments in Gravitational Physics* (1993) Cambridge Univ. Press.

[136] G.W. Gibbons and B.F. Whiting, *Nature* 291 (1981) 636.

[137] E. Fischbach, D. Sudarsky, A. Szafer, C. Talmadge, S. Aroson, *Phys. Rev. Lett.* 56, 3, 1986.

[138] C.C. Speake and T.J. Quinn, *Phys. Rev. Lett.* 61, 1340 (1988).

[139] D. Eckhardt, C. Jekeli, A. Lazarewicz, A. Romaides, *Phys. Rev. Lett.* 60, 2567 (1988).

[140] D.H. Eckhardt, *Phys. Rev.* 48 D (1993) 3762.

[141] C. Misner, K. Thorne, J. Wheeler, Gravitation, W. Freeman and Company, N Y (1970).

[142] C. Lanczos, *The Variational Principles of Mechanics*, Toronto Univ. Press (1949).

[143] E. Schwinger, *Phys. Rev.* 82, 914 (1951), *Phys. Rev.* 91, 713 (1953).

[144] E.V. Ferapontov, M.V. Pavlov, arXiv:nlin.SI/0212026 (2002).

[145] G. Basini and S. Capozziello, *Gen. Rel. Grav* 37, 115 (2005).

[146] F. Reines and C.L. Cowan Jr., *Phys. Rev.* 92, 830 (1953).

[147] J.I. Collar, *Phys. Rev. Lett.* 76, 999 (1996).

[148] S. Capozziello, G. Lambiase, G. Iovane, *Mod. Phys. Lett. A* 18, 905 (2003).

[149] M. Blasone, A. Capolupo, S. Capozziello, S.Carloni, G. Vitiello, *Phys. Lett. A* 323, 182 (2004).

[150] G.L. Fogli *et al.*, hep-ph/0408045 (2004).

[151] D. Ahluwalia, C. Burgard, *Gen. Rel. Grav.* 28, 1161(1996) *Phys. Rev. D* 57, 4724 (1998).

[152] Y. Grossman and H.J. Lipkin, *Phys. Rev. D*55, 2760 (1997).

[153] K. Konno and M. Kasai, *Progr. Theor. Phys.* 100, 1145 (1998).

[154] T. Bhattacharya, S. Habib and E. Mottola, *Phys. Rev.* D59, 067301 (1999).

[155] J.D. Anderson, et al., *Phys. Rev. D* 65 (2002) 082004.

[156] D. La, P.J. Steinhard, and E.W. Bertschinger, *Phys. Lett.* B220, 375 (1989).

[157] D. La and P.J. Steinhard, *Phys. Rev. Lett.* 62, 376 (1989).

[158] R.Pound and G.A. Rebka, *Phys. Rev. Lett.* 4, 337 (1960).

[159] R. Colella, A.W. Overhauser, S.A. Werner, *Phys. Rev. Lett.* 34, 1472 (1975).

[160] http://www.ligo.org/pdf public/camp.pdf. B. Abbott et al., LIGO: the Laser Interferometer Gravitational-Wave Observatory, *Rep. Prog. Phys.* 72, 076901-50, (2009), S. Finn et al., LIGO Doc. ligo-t1000251, The LIGO Scientific Collaboration (2010).

[161] http://www.ligo.org/pdf public/hough02.pdf, F. Acernese et al., The present status of the VIRGO Central Interferometer, *Class. Quant. Grav.* 21, (2004), 385-394.
H. Grote (LIGO Coll.), Status GEO600, *Class. Quantum Grav.* 25, (2008), 114043. K. Kuroda and LCGT Coll., Status LCGT, *Class. Quantum Grav.* 27, (2010),

[162] S. Capozziello, A. Troisi, *Phys. Rev. D*72, 044022 (2005)

[163] S. Capozziello - Newtonian Limit of Extended Theories of Gravity in *Quantum Gravity Research Trends* Ed.;
A. Reimer, pp. 227-276 Nova Science Publishers Inc., NY (2005).

[164] M. E. Tobar, T. Suzuki and K. Kuroda, *Phys. Rev. D* 59, 102002 (1999).

[165] B. Allen-Proc. of the *Les Houches School on Astrophysical Sources of Gravitational Waves*, Ed. J.A. Marck, J.P. Lasota (Cambridge Univ. Press, 1998).

[166] B. Allen and A.C. Ottewill - *Phys. Rev. D* 56, 545-563 (1997).

[167] M. Maggiore - *Phys. Rep.* 331, 283-367 (2000).

[168] L. Grishchuk et al. - *Phys. Usp.* 44 1-51 (2001); *Usp. Fiz. Nauk* 171 3-59 (2001).

[169] C.L. Bennet et al. - *ApJS* 148, 1 (2003).

[170] D.N. Spergel et al. - *ApJS* 148, 195 (2003).

[171] S. Nojiri and S.D. Odintsov, *Int. J. Geom. Meth. Mod. Phys.* 4, 115 (2007).

[172] C. Brans and R. H. Dicke - *Phys. Rev.* 124, 925 (1961).

[173] B. Allen - *Phys. Rev.* D 3-7, 2078 (1988).

[174] G.Watson *"An exposition on inflationary cosmology"* N. Carol. Univ. Press (2000).

[175] A. Guth - *Phys. Rev.* 23, 347 (1981).
S. Capozziello, Ch. Corda, M. De Laurentis, *Mod. Phys. Lett.* A 22, 1097 (2007).

[176] G. Basini, S. Capozziello, *General Relativity & Gravitation*, V. 38, March (2006).

[177] S. Capozziello, G. Basini, M. DeLaurentis, *Europ. J. of Phys.* C 71:1679, (2011).

[178] G. Basini, S. Capozziello, M. DeLaurentis, *Modern Phys. Lett. A V.* 27,(2012).

[179] G.Basini, S. Capozziello, M. DeLaurentis, *Int. J. Geom. Meth. Mod. Phys. WPS*, 2015.

[180] G.Basini, S. Capozziello, M. De Laurentis, *Int. J. Geom. Meth. Mod. Phys.* (2015).

In: New Research on Quantum Entanglement ISBN: 978-1-63482-888-8
Editor: Lori Watson, pp. 99-116

Chapter 2

ENTANGLEMENT RELATIVITY IN THE FOUNDATIONS OF THE OPEN QUANTUM SYSTEMS THEORY

M. Arsenijević[1,*]***, J. Jeknić-Dugić***[2]***, D. Todorović***[1] ***and M. Dugić***[1]
[1]Department of Physics, Faculty of Science,
Kragujevac, Serbia
[2]Department of Physics, Faculty of Science and Mathematics,
Niš, Serbia

Abstract

Realistic many-particle systems dynamically exchange particles with their environments. In classical physics, small variations in the number of constituent particles are commonly considered practically irrelevant. However, in the quantum mechanical context, such and similar structural variations are generically taxed due to the so-called Entanglement Relativity. In this paper we point out difficulties in deriving master equation for a subsystem of an alternative partition of the closed quantum system. We find that the Nakajima-Zwanzig projection method cannot be straightforwardly used to solve the problem. The emerging tasks and prospects for the consistent foundations are examined.

PACS: 03.65.Ud, 03.65.Yz, 03.65.Ta

*E-mail address: fajnman@gmail.com

1. Introduction

Realistic many-particle systems dynamically exchange particles with their environments and with other systems. This trivial observation is still largely intact in the foundations of quantum theory, including quantum measurement, decoherence and the open quantum systems theory. While intuitively exchange of particles is completely clear, it sets a blurred separating line between a system and its environment. Classical physics straightforwardly tackles such situations e.g., in motion with variable mass or in the Grand Canonical Ensemble for systems in thermal equilibrium. However, systematic quantum-mechanical description of such processes is lacking.

In quantum measurement or decoherence [1, 2, 3] or more generally in open quantum systems [4, 5, 6] theory, the border line between 'system' and 'environment' is rarely analyzed in depth. We construe this fact as a symptom of a subtle and hard problem that is the subject of this paper.

The aim of this paper is to diagnose the problem that is still largely unrecognized but nevertheless of the fundamental importance in the field of open quantum systems and applications. We hope that, this first step in noticing and identifying the problem can help in setting outlines of further progress in the field.

In Section 2 we carefully define the problem. In Section 3 we generalize our findings in the context of the so-called projection-method approach. Section 4 is an illustration of our considerations and their subtlety. Section 5 is conclusion.

2. The Problem

Planet Earth is constantly bombarded from the outer space. Provided that the captured-projectile mass is much less than the mass of the Earth, the variations in the Earth's orbit around the Sun are practically negligible. Classical statistical physics routinely accounts for huge stochastic change in the number of particles in the many-particle systems in thermal equilibrium. The Grand Canonical Ensemble of the standard classical statistical mechanics describes a composite system of N particles with the variations of the number of particles from the set $\{0, 1, 2, 3, ..., N\}$. There is also no problem with description of non-equilibriun mesoscopic systems, such as Brownian particle, which is dressed by the water molecules that constantly stick to and come off the particle's surface–the particle's 'hydration shell' known also for large molecules (e.g., protein molecules and other biopolymers) in a solution.

Except for the 'chaotic' systems, it seems that classical physics embraces the following rule: Knowledge of dynamics of an N-particle classical system allows *straightforward deduction* of dynamics for the $(N \pm n)$-particle systems under the same physical conditions (external fields and interactions), as long as $n \ll N$. More intuitively, it seems that individuality of many-particle classical systems is not threatened by tiny changes in the system's separation from the rest of the world–e.g., Brownian particle with n attached water molecules is typically regarded as practically the same system as with n' ($\neq n$) attached water molecules.

However, in the quantum mechanical context, the things stand differently. To see this, let us consider the Hamiltonian of interest in the quantum Brownian model, in which the environment E monitors the particle's center of mass position [4] (and references therein). For simplicity, we refer to one dimensional system and the environment consisting of non-interacting linear harmonic oscillators:

$$\hat{H} = \hat{H}_S + \hat{H}_E + \hat{H}_{SE} \tag{1}$$

where

$$\hat{H}_S = \sum_{i=1}^{N_S} \frac{\hat{p}_i^2}{2m_i} + \sum_{i\neq j=1}^{N_S} V(|\hat{x}_i - \hat{x}_j|)$$
$$\hat{H}_E = \sum_{\alpha=1}^{N_E} \left(\frac{\hat{p}_\alpha^2}{2m_\alpha} + \frac{1}{2} m_\alpha \omega_\alpha^2 \hat{x}_\alpha^2 \right)$$
$$\hat{H}_{SE} = \hat{X}_{CM} \otimes \sum_{\alpha=1}^{N_E} \kappa_\alpha \hat{x}_\alpha \tag{2}$$

where the Latin indices refer to the S system and the Greek indices to the E system, with the numbers N_S and N_E of particles in the system and the environment, respectively, with the pair interactions V, while $\hat{X}_{CM} = \sum_{i=1}^{N_S} m_i \hat{x}_i / M$ and $M = \sum_{i=1}^{N_S} m_i$.

Consider now that the $i_\circ$th particle of the S system becomes a part of the environment. This is, instead of the S system and the environment E there is the new open system $S' = S \setminus i_\circ$ and the new environment $E' = E \cup i_\circ$ which symbolically reads as a structural (not necessarily dynamical) transformation:

$$\{S, E\} \to \{S', E'\}, \tag{3}$$

while the composite system C as a whole remains intact by the transformation: $S + E = C = S' + E'$.

Needless to say, the composite system's Hamiltonian eq.(1) remains intact by the transformation, while it takes another form:

$$\hat{H} = \hat{H}_{S'} + \hat{H}_{E'} + \hat{H}_{S'E'} \tag{4}$$

where:

$$\hat{H}_{S'} = \sum_{i=1}^{N_S-1} \frac{\hat{p}_i^2}{2m_i} + \sum_{i\neq j=1}^{N_S-1} V(|\hat{x}_i - \hat{x}_j|)$$
$$\hat{H}_{E'} = \hat{H}_E + \frac{\hat{p}_{i_\circ}^2}{2m_{i_\circ}} + \frac{1}{M}\hat{x}_{i_\circ} \otimes \sum_{\alpha=1}^{N_E} \kappa_\alpha \hat{x}_\alpha$$
$$\hat{H}_{S'E'} = \frac{1}{M} \sum_{i=1}^{N_S-1} m_i \hat{x}_i \otimes \sum_{\alpha=1}^{N_E} \kappa_\alpha \hat{x}_\alpha + \sum_{j=1}^{N_S-1} V(|\hat{x}_{i_\circ} - \hat{x}_j|). \tag{5}$$

The following simplifications can make the two models eq.(2) and eq.(5) similar to each other: (i) for $N_S \gg 1$, the total mass $M \approx M' = \sum_{i=1}^{N_S-1} m_i$, (ii) for large M (i.e. M'), one can neglect the last term in $\hat{H}_{E'}$, or more generally to introduce the normal coordinates for the new environment E' such that it consists of mutually non-interacting quasi-particles [7, 8] and (iii) the pair interactions in $\hat{H}_{S'E'}$ to consider as a weak perturbation. Then, according to the classical intuition, the two open systems S and S' and their dynamics should appear essentially mutually indistinguishable.

However, formal similarity of the quantum Hamiltonians does not in general guarantee that reduced dynamics of the S system straightforwardly implies the reduced dynamics of the S' system. This subtle point requires careful examination.

Central to derivation of master equations in the density matrix theory of open quantum systems is the tensor product:

$$\hat{\rho}_S(t) \otimes \hat{\rho}_E, \tag{6}$$

which is an *ansatz* known as *Born approximation* [4, 5, 6] and follows in a systematic way from the projection of the composite system's state $\hat{\rho}(t)$ [9, 10]:

$$\mathcal{P}\hat{\rho}(t) = \hat{\rho}_S(t) \otimes \hat{\rho}_E, \tag{7}$$

with $\mathcal{P}^2 = \mathcal{P}$ and $\mathcal{Q} = \mathcal{I} - \mathcal{P}$ such that $\mathcal{Q}^2 = \mathcal{Q}$, while $\hat{\rho}_S(t) = tr_E \hat{\rho}(t)$ carries all information about the open system S. Linearity of $\mathcal{P}$ excludes the choice $\hat{\rho}_E = tr_S \hat{\rho}(t)$.

Let us suppose that the reduced dynamics for the S system is well described by a proper master equation, which assumes validity of eq.(6), i.e., eq.(7). The main observation of this paper is as follows:

($\mathcal{O}$) *As distinct from the classical counterpart, reduced dynamics of the S system cannot in general be used to derive or deduce dynamics of the S' system for the same time interval.*

This observation asserts that a master equation for the S system cannot be used to deduce/derive master equation for the S' system for the same time interval, $[0, t]$, with the fixed initial state, $\hat{\rho}(t = 0)$, of the total C system.

In order to justify the ($\mathcal{O}$), we first emphasize the formal yet substantial distinction between the classical and quantum state spaces. In classical physics, the 'phase space' of N particles is Cartesian product of the N 'phase spaces' for individual particles. However, in quantum mechanics, the particles state spaces are not in Cartesian but in tensor product. For the composite system C decomposed (structured) as $S + E$, the Hilbert state space $\mathcal{H}$ is the tensor product of the state spaces for the subsystems:

$$\mathcal{H} = \mathcal{H}_S \otimes \mathcal{H}_E. \tag{8}$$

The structural transformation eq. (3) gives rise to re-factorization:

$$\mathcal{H} = \mathcal{H}_{S'} \otimes \mathcal{H}_{E'}. \tag{9}$$

Invariants of the transformation eq. (3) are: (a) the Hilbert state space $\mathcal{H}$, (b) the composite system's Hamiltonian $\hat{H}$ and (c) the composite system's state in every instant of time. However, the transformation eq.(3) induces a change in the form of the Hamiltonian (e.g., eqs. (2) and (5)) as well as re-factorization eq.(8) $\to$ eq.(9). The transformation eq.(3) also leads to a change *in the form* of the composite system's instantaneous state. For a pure state in an instant of time t, this is known as *Entanglement Relativity*, e.g., the equality:

$$|\phi\rangle_S \otimes |\chi\rangle_E = \sum_i c_i |i\rangle_{S'} \otimes |i\rangle_{E'}. \tag{10}$$

Entanglement Relativity (ER) is a recently established (and rediscovered) [11 - 16] corollary of quantum mechanics that asserts: Virtually every structural transformation that induces tensor re-factorization also induces a change in amount of quantum entanglement in the composite system C. If the composite system's state is tensor-product for one structure $(S + E)$ it is of the entangled form for practically all[1] the alternative structures $(S' + E')$ of the composite system. ER regards every (pure) state in every instant in time t.

On the basis of ER, it can be shown that also the more general non-classical correlations, quantified e.g., by 'quantum discord', are structure-dependent [17]: A tensor-product mixed state for one structure $(S + E)$ acquires a *quantum* correlated (entangled or discordant) form for virtually arbitrary alternative structure $(S' + E')$ of the composite system, e.g.:

$$\hat{\rho}_S \otimes \hat{\rho}_E = \sum_i \lambda_i \hat{\rho}_{S'i} \otimes \hat{\rho}_{E'i}, \quad \sum_i \lambda_i = 1. \tag{11}$$

Now the classically unknown conditions eqs. (10) and (11) plausibly justify the $(\mathcal{O})$ statement as follows; the more rigorous consideration is given in Section 3. The projection eq.(7), as well as the generalizations given in Section 3, do not possess any quantum correlations. Equations (6) and (7) are of exactly the same form as the l.h.s. of eqs. (10) and (11), which refer to the $S+E$ structure. The r.h.s. of eqs. (10) and (11), which refer to the $S' + E'$ structure, carry the quantum correlations that are not accounted for by any generalization of eq.(7). Hence derivation of master equation that is *based* on eq.(6), i.e., on eq.(7), for the S system is practically never a simultaneous derivation of the master equation for the S' system. Consequently, the knowledge of dynamics of the S system does not suffice to conclude much about dynamics of the S' system.

For the probably most relevant class of Markovian open systems, eqs. (10) and (11) reveal another layer of consideration. The tensor-product initial state $\hat{\rho}_S(t = 0) \otimes \hat{\rho}_E$ is a necessary condition for Markovian dynamics [5, 6]; typically, the environment is supposed thermal, $\hat{\rho}_E = \exp(-\beta \hat{H}_E)/tr_E \exp(-\beta \hat{H}_E)$, on the inverse temperature $\beta = 1/k_B T$. Due to eqs. (10) and (11) for $t = 0$, as long as eq.(6) is valid for the $S + E$ structure, it is practically never fulfilled regarding the alternative $S' + E'$ structure. Then Markovian dynamics for the S system is virtually never applicable for the alternative S' system. More specifically: even if eq.(5) can be reduced to eq.(2), and even if the new environment may also be in thermal-equilibrium

[1] Exceptions to ER are also known, but do not alter our main point. e.g., the transformation eq.(3) does not change a tensor-product state. However, the more general structural transformations change even such states and ER applies [16].

state, $\sum_i \lambda_i \hat{\rho}_{S'i} = \exp(-\beta' \hat{H}_{E'})/tr_E(\exp(-\beta' \hat{H}_{E'}))$, there is initial correlation for the S' and E' systems. Consequently, the reduced S' system's dynamics is not Markovian [5, 6] and is also possibly non-completely positive [18].

Hence, typically, derivation of the S' system's dynamics has to be started from the scratch–by setting eq.(6) for the $S' + E'$ structure in an independent derivation of master equation for the S' system. To this end, the knowledge about dynamics of the S system is not useful.

3. On the Use of the Nakajima-Zwanzig Projection Method

One may still wonder if, somehow, dynamics of the S system can be used for drawing conclusions on the S' system's dynamics. After all, it's just one tiny-particle difference between the two structures pertaining to eq. (2) and eq. (5). The correlations present on the r.h.s. of eqs. (10) and (11) are due only to a single particle denoted $i_\circ$. May it be possible to approximate the r.h.s. of eqs. (10) and (11) by some tensor-product states?

In certain special cases (e.g., when the use of the Born-Oppenheimer approximation is allowed), this may be the case. Nevertheless there still remains open the question as to whether 'tiny correlations' can be safely discarded from considerations in general. In this paper we are not interested in such subtle and deep questions. Rather, we consider usefulness of the Nakajima-Zwanzig projection method [9, 10] in regard of the ($\mathcal{O}$) statement.

Our motivation comes from the fact that the Nakajima-Zwanzig and the related (projection-based) methods provide systematic introduction of eq.(6) and set the basis for the up-to-date the most general methodological basis of the open systems field [4, 5, 6]. If the ($\mathcal{O}$) statement remains valid in the context of the projection-based methods, then it presents a serious limitation not only to our classical intuition but also to operational procedures in describing dynamics of the alternate open systems. Bearing in mind the classical intuition of Sections 1 and 2, in such a case we face yet another non-trivial task in the context of the problem of transition from quantum to classical.

Below, we assume arbitrary bipartitions of a composite system C, $S + E = C = S' + E'$, i.e., arbitrary linear canonical transformations (LCTs) that induce the tensor-product structures (i.e., tensor re-factorization) of the composite system's Hilbert state space.

The key idea behind the Nakajima-Zwanzig projection method [9, 10] is presented by eq.(7). It consists of the introduction of a linear projection operator, $\mathcal{P}$, which acts on the operators of the state space of the composite system 'system+environment' ($S + E$). If $\hat{\rho}$ is the density matrix of the composite system, the projection $\mathcal{P}\hat{\rho}$ (the 'relevant part' of the composite density matrix) serves to represent a simplified effective description through a reduced state of the composite system. The complementary part (the 'irrelevant part' of the composite density matrix), $\mathcal{Q}\hat{\rho} = (I - \mathcal{P})\hat{\rho}$. For the 'relevant part', $\mathcal{P}\hat{\rho}(t)$, one derives closed ('autonomous') equations of motion in the form of integro-differential equation. The open system's density matrix $\hat{\rho}_S(t) = tr_E \mathcal{P}\hat{\rho}(t)$ is *required* to carry *all* information about the open system S, equivalently $tr_E \mathcal{Q}\hat{\rho} = 0$.

The Nakajima-Zwanzig projection method assumes a concrete, in advance chosen and fixed for all time-instants, system-environment split (a 'structure'), $S+E$. This split is uniquely defined by the associated tensor product structure of the composite system's Hilbert space, $\mathcal{H} = \mathcal{H}_S \otimes \mathcal{H}_E$. Division of the composite system into 'system' and 'environment' is practically motivated. In principle, the projection method can equally describe arbitrary system-environment split i.e., arbitrary factorization of the composite system's Hilbert space. By definition, different factorizations introduce different projectors, denoted $\mathcal{P}$ for the $S + E$ structure, and $\mathcal{P}'$ for some alternative $S' + E'$ structure of the composite system, such that $\hat{\rho}_S = tr_E \mathcal{P}\hat{\rho}(t)$ carries all information about the S system (equivalently $tr_E \mathcal{Q}\hat{\rho}(t) = 0$), and $\hat{\rho}_{S'} = tr_{E'} \mathcal{P}'\hat{\rho}(t)$ carries all information about the S' system (equivalently $tr_{E'} \mathcal{Q}'\hat{\rho}(t) = 0$).

The linear projections can be defined [1, 19]: (i) $\mathcal{P}\hat{\rho}(t) = (tr_E \hat{\rho}(t)) \otimes \hat{\rho}_E$ [for some $\hat{\rho}_E \neq tr_S \hat{\rho}$], which is eq.(7), (ii) $\mathcal{P}\hat{\rho}(t) = \sum_n (tr_E \hat{P}_{Sn} \hat{\rho}(t)) \otimes \hat{\rho}_{En}$ [with arbitrary orthogonal supports for the $\hat{\rho}_E$s], and (iii) $\mathcal{P}\hat{\rho}(t) = \sum_i (tr_E \hat{P}_{Ei} \hat{\rho}(t)) \otimes \hat{P}_{Ei}$ [with arbitrary orthogonal projectors for the E system]; by $\hat{P}$, we denote the projectors on the respective Hilbert state (factor) spaces. The physical context fixes the choice of the projection–e.g., by an assumption about the initial state. In this paper we stick to the projection (i), which is by far of the largest interest in foundations and applications of the open systems theory. As it can be easily shown, all the projections (i)-(iii) are free from the quantum correlations (entanglement or discord).

Now we provide the main results of this section that are borrowed from [20] with the proofs placed in the appendices.

Lemma 1. *For the most part of the composite system's dynamics, validity of*

$$tr_E \mathcal{Q}\rho(t) = tr_E(\hat{\rho}(t) - \mathcal{P}\hat{\rho}(t)) = tr_E(\hat{\rho}(t) - \hat{\rho}_S(t) \otimes \hat{\rho}_E) = 0, \forall t. \quad (12)$$

implies non-validity of

$$tr_{E'} \mathcal{Q}\hat{\rho}(t) = tr_{E'}(\hat{\rho}(t) - \rho_S(t) \otimes \hat{\rho}_E) = 0, \forall t, \quad (13)$$

and vice versa.

Lemma 1 reveals that the information 'irrelevant part' of a projected state for one structure contains some relevant information regarding an alternative structure of the composite system for the most of time instants t. In formal terms: for the most part of the composite system's dynamics, the projection $\mathcal{Q}\hat{\rho}$ ($\mathcal{Q}'\hat{\rho}$) brings some information about the open system S' (S)–at variance with the Nakajima-Zwanzig projection idea. Hence $\partial\mathcal{P}\hat{\rho}(t)/\partial t$ allows 'tracing out' regarding only one structure. If that structure is $S + E$, then $tr_{E'}\partial\mathcal{P}\hat{\rho}(t)/\partial t \neq \partial\hat{\rho}_{S'}(t)/\partial t$ [as long as $\hat{\rho}_{S'}(t) = tr_{E'}\hat{\rho}(t)$]. This can be seen also from the following argument, which is not restricted to the projection-based methods. Tracing out the E system is dependent on, but not equal to, the tracing out the E' system, and *vice versa*. This dependence follows from the fact that the S and E degrees of freedom are intertwined with the S' and E' degrees of freedom. Intuitively: 'tr_E' (e.g., integrating over the E's degrees of freedom) *partly* encompasses both the S' and the E' degrees of freedom.

Lemma 2. *The two structure-adapted projectors $\mathcal{P}$ and $\mathcal{P}'$ do not mutually commute*

$$[\mathcal{P}, \mathcal{P}']\hat{\rho}(t) \neq 0 \quad (14)$$

for the most of the time instants t.

Very much like noncomutativity of quantum observables, Lemma 2 asserts that the projection-based information contents regarding different structures of a composite system are mutually *exclusive* for the most of the time instants t. Formally, there is no state $\hat{\rho}(t)$ of a composite system for which the equality $\mathcal{P}\hat{\rho}(t) = \hat{\rho}(t) = \mathcal{P}'\hat{\rho}(t)$ can be fulfilled for arbitrary instant of time t.

Lemma 1 and lemma 2 refer to all projection-based methods and exclude acquisition of information about an open system S' from the master equation known for the alternative (albeit possibly similar) open system S in the same instant (or interval) of time, and *vice versa*. In effect, the ($\mathcal{O}$) statement is justified and leads to the conclusion that derivation of master equations has to be performed for every set of the degrees of freedom (i.e., for every open system, S, S' etc.) separately, in accord with equations (10) and (11).

4. Analysis of the Quantum Brownian Motion

In order to illustrate subtlety of our considerations, we stick to eq.(2) as the Caldeira-Leggett model of quantum Brownian motion (QBM) [4, 21].

In the Schrödinger picture the QBM master equation for the initial separable state ($\hat{\rho}(t=0) = \hat{\rho}_S(t=0) \otimes \hat{\rho}_E$) with the environment on temperature T:

$$\frac{d\hat{\rho}_S(t)}{dt} = -\frac{\imath}{\hbar}[\hat{H}_S, \hat{\rho}_S(t)] - \frac{\imath\gamma}{\hbar}[\hat{x}_S, \{\hat{p}_S, \hat{\rho}_S(t)\}] - \frac{2m\gamma k_B T}{\hbar^2}[\hat{x}_S, [\hat{x}_S, \hat{\rho}_S(t)]]. \quad (15)$$

The curly brackets denote the 'anticommutator', m is the mass while $\hat{x}_S$ and $\hat{p}_S$ are the position and momentum of the particle and γ is the semi-empirical friction coefficient.

Eq.(15) is not of the Lindblad form and hence by definition [5, 6] is not Markovian. Interestingly enough, eq.(15) applies even for initially correlated state and for arbitrary strength of interaction in the composite system as well as for arbitrary 'spectral density' (which defines the friction coefficient γ)[2]. Non-Markovianity of eq.(15) is behind its 'robustness', which is the ultimate basis of the observation of QBM effect for an alternative structure of the total system C [7].

Now we emphasize the variations offered by eq.(15). First, it is known that eq.(15) can be transformed in a Lindblad form for sufficiently high temperature T [4]. Second, for the massive particle, the second term proportional to γ can be neglected. This constitutes the 'recoilless' variant of the Caldeira-Leggett model and provides the Lindblad-form master equation:

$$\frac{d\hat{\rho}_S(t)}{dt} = -\frac{\imath}{\hbar}[\hat{H}_S, \hat{\rho}_S(t)] - \frac{2m\gamma k_B T}{\hbar^2}[\hat{x}_S, [\hat{x}_S, \hat{\rho}_S(t)]], \quad (16)$$

which is similar with the scattering-decoherence master equation–see eq.(3.66) in [1].

It is remarkable that already at the level of *fixed structure*, $C = S + E$, we can see non-trivial variations in the form of master equation and consequently regarding the S system's dynamics.

[2]See e.g., eq.(4.226) in Ref. [4].

Now we refer to certain *structural variations* of the dynamics described by eq.(16). We are aiming at the cases in which the classical intuition can be justified; see e.g., comments below eq.(5). In all other cases we do not expect the classical intuition to be very useful.

Concretely, we are interested in eq.(3) as well as in the opposite case, i.e., when an environmental particle, denoted $\alpha_\circ$, is joined the S system thus providing a new open system, $S'' = S \cup \alpha_\circ$, and new environment, $E'' = E \setminus \alpha_\circ$; needless to say, $C = S + E = S' + E' = S'' + E''$.[3] This situation also describes the 'Schrödinger's cat'–the cat represented by the S system while the $\alpha_\circ$th environmental particle flows out of the radioactive source.

Regarding the structural transformation

$$(S, E) \to (S'', E''), \tag{17}$$

that is accompanied by tensor re-factorization, $\mathcal{H}_S \otimes \mathcal{H}_E \to \mathcal{H}_{S''} \otimes \mathcal{H}_{E''}$, the Hamiltonian eq.(2) i.e., eq.(5) takes the form:

$$\hat{H} = \hat{H}_{S''} + \hat{H}_{E''} + \hat{H}_{S''E''}. \tag{18}$$

In eq.(18):

$$\hat{H}_{S''} = \hat{H}_S + \frac{\hat{p}^2_{\alpha_\circ}}{2m_{\alpha_\circ}} + \hat{X}_{CM} \otimes \kappa_{\alpha_\circ} \hat{x}_{\alpha_\circ}$$
$$\hat{H}_{E'} = \sum_{\alpha=1}^{N_E - 1} \left(\frac{\hat{p}^2_\alpha}{2m_\alpha} + \frac{1}{2} m_\alpha \omega^2_\alpha \hat{x}^2_\alpha \right)$$
$$\hat{H}_{S''E''} = \hat{X}_{CM} \otimes \sum_{\alpha=1}^{N_E - 1} \kappa_\alpha \hat{x}_\alpha. \tag{19}$$

As distinct from eq.(5), eq.(19) is already of the form of eq.(2): there only appears a new interaction in the system's (in the S'' system's) self Hamiltonian[4]. Therefore, the two structure variations, eq.(3) and eq.(17), are not mutually equivalent.

[3]For examples of the more general structural transformations see e.g., Refs. [7, 16, 22, 23].

[4]Rigorously, the new structure $(S + \alpha_\circ) + E''$, where the $\alpha_\circ$ particle is not in interaction with E''. If we assume that the mass M'' of S'' is approximately M, the two models eq.(2) and eq.(19) become practically indistinguishable.

Our starting model is eq.(2), for which we assume the initial tensor product state and thermal environment E:

$$\hat{\rho}(t=0) = \hat{\rho}_S \otimes \hat{\rho}_E = \hat{\rho}_S \otimes_\alpha \frac{\exp(-\beta \hat{H}_\alpha)}{Z_\alpha}, \tag{20}$$

with the one-particle 'statistical sum' $Z_\alpha = tr_\alpha \exp(-\beta \hat{H}_\alpha)$.

The interactions generated by the V pair-interactions in the S system suggest correlations in the initial S system's state, $\hat{\rho}_S(t=0) = \sum \mu_m \hat{\rho}_m^{S'} \otimes \hat{\rho}_m^{i_\circ}$, which gives correlated initial state for the $S' + E'$ structure:

$$\hat{\rho}(t=0) = \sum_m \mu_m \hat{\rho}_m^{S'} \otimes \hat{\rho}_m^{E'} \equiv \sum_m \mu_m \hat{\rho}_m^{S'} \otimes \left(\hat{\rho}_m^{i_\circ} \otimes \hat{\rho}^E\right), \quad \sum_m \mu_m = 1. \tag{21}$$

However, at variance with eq.(11)[5], eq.(20) directly provides tensor-product initial state for the $S'' + E''$ structure:

$$\hat{\rho}(t=0) = \hat{\rho}_{S''} \otimes \hat{\rho}_{E''} \equiv \left(\hat{\rho}_S \otimes \frac{\exp(-\beta \hat{H}_{\alpha_\circ})}{Z_{\alpha_\circ}}\right) \otimes_{\alpha \neq \alpha_\circ} \frac{\exp(-\beta \hat{H}_\alpha)}{Z_\alpha}. \tag{22}$$

Further we focus on the $S'' + E''$ structure.

The projection is defined:

$$\mathcal{P}\hat{\rho}(t) = \hat{\rho}_S(t) \otimes_\alpha \frac{\exp(-\beta \hat{H}_\alpha)}{Z_\alpha}, \tag{23}$$

providing that $\otimes_\alpha \exp(-\beta \hat{H}_\alpha)/Z_\alpha \neq tr_S \hat{\rho}(t)$ (and analogously for the other structures).

For the $S'' + E''$ structure, the projection reads:

$$\mathcal{P}''\hat{\rho}(t) = \left(tr_{E\setminus\alpha_\circ}\hat{\rho}(t)\right) \otimes \hat{\rho}_{E''}. \tag{24}$$

In an instant of time $t > 0$, we expect correlations in the S'' system, e.g., $\hat{\rho}_{S''}(t) := tr_{E\setminus\alpha_\circ}\hat{\rho}(t) = \sum_i \lambda_i(t) \hat{\rho}_i^S(t) \otimes \hat{\rho}_i^{\alpha_\circ}(t)$. Hence with the use of eq.(24):

$$tr_{E''} \frac{\partial \mathcal{P}''\hat{\rho}(t)}{\partial t} = \frac{\partial}{\partial t} \sum_i \lambda_i(t) \hat{\rho}_i^S(t) \otimes \hat{\rho}_i^{\alpha_\circ}(t). \tag{25}$$

[5]This nicely exhibits the subtlety of ' quantum correlations relativity', eq.(10) and eq.(11): for some special states (here: tensor-product states) and for a special pair of structures (here: $S + E$ and $S'' + E''$) one should not worry about the quantum correlations relativity. However, the worry remains for almost all other kinds of re-structuring even for the initial tensor-product state regarding the starting $S + E$ structure of the total system.

On the other hand, from eq.(23):

$$tr_{E''}\mathcal{P}\hat{\rho}(t) = \hat{\rho}_S(t) \otimes \frac{\exp(-\beta \hat{H}_{\alpha_\circ})}{Z_{\alpha_\circ}}, \qquad (26)$$

which instead of eq.(25) gives:

$$tr_{E''}\frac{\partial \mathcal{P}\hat{\rho}(t)}{\partial t} = \frac{\partial}{\partial t}\hat{\rho}_S(t) \otimes \frac{\exp(-\beta \hat{H}_{\alpha_\circ})}{Z_{\alpha_\circ}}. \qquad (27)$$

The absence of the exact correlations appearing in eq.(25) clearly illustrates Lemma 1, i.e., implies that $tr_{E''}\mathcal{Q}\hat{\rho}(t) \neq 0$, even in this case[6] in which eq.(11) is not applicable.

Hence despite the classical similarity, the open systems S, S' and S'' are subjected to different dynamics. While the S system, as well possibly as the S'' system, undergoes a frictionless Markovian dynamics eq.(16), the S' system may be expected to be subjected to neither Markovian [5, 6] nor frictionless [4, 21] and possibly non-completely-positive [18] dynamics. Exact master equations for the S' and S'' systems can follow from the independent e.g., projection-based analysis. Regarding the S' system, derivation of the master equation should start from the projection $\mathcal{P}'\hat{\rho}(t) = \hat{\rho}_{S'}(t) \otimes \hat{\rho}_{E'}$ with the initial state of the form of eq.(21), while for the S'' system, the derivation should regard eq.(25) with the initial state of the form of eq.(22)–that does not bring any substantial new observation that is of interest for the present paper.

5. Conclusion

Recently, it was shown [7] that non-similar subsystems of a composite system can have similar dynamics. In contrast to the classical intuition of Sections 1 and 2, in this paper we show that classically indistinguishable many-particle systems can undergo non-equivalent dynamics. Moreover, the shortcuts in describing dynamics of the alternative open systems may not even exist. It is our conjecture that the classical intuition fits with some special structures of many-particle systems [24] that can still require certain assumptions, such as e.g., the Born-Oppenheimer adiabatic approximation. Those results set a new layer in the long standing problem of the transition from quantum to classical [1] that will be discussed elsewhere.

[6]See eq.(22).

Perhaps not surprisingly, our findings are implied by the classically unknown Quantum Correlations Relativity [11 - 17]–a not-yet-fully-appreciated rule of the universally valid quantum theory–and some other, classically surprising, findings may be expected.

Currently it appears that description of the alternative subsystems dynamics should be performed for every alternate open system separately. The classical intuition, that similar systems should bear similar dynamics, appears unreliable.

Acknowledgment

The work on this paper is financially supported by Ministry of education, science and technological development, Serbia, grant no 171028 and in part for MD by the ICTP-SEENET-MTP grant PRJ-09 "Strings and Cosmology" in frame of the SEENET-MTP Network.

Appendix A. Proof of Lemma 1

Given eq.(12), i.e., $tr_E \mathcal{Q}\hat{\rho}(t) = 0, \forall t$, we investigate the conditions that should be fulfilled in order for eq.(13), i.e., $tr_{E'} \mathcal{Q}\hat{\rho}(t) = 0, \forall t$, to be fulfilled. The $\mathcal{Q}$ projector refers to the $S + E$, not to the $S' + E'$ structure. Therefore, in order to calculate $tr_{E'} \mathcal{Q}\hat{\rho}(t)$, we use ER. We refer to the projection (i), Section 3, in an instant of time:

$$\mathcal{P}\hat{\rho} = (tr_E\hat{\rho}) \otimes \hat{\rho}_E. \tag{28}$$

A) Pure state $\hat{\rho} = |\Psi\rangle\langle\Psi|$, while, due to eq.(12), $tr_E \mathcal{Q}|\Psi\rangle\langle\Psi| = 0$.

We consider the pure state presented in its (not necessarily unique) Schmidt form

$$|\Psi\rangle = \sum_i c_i |i\rangle_S |i\rangle_E, \tag{29}$$

where $\hat{\rho}_S = tr_E|\Psi\rangle\langle\Psi| = \sum_i p_i |i\rangle_S\langle i|$, $p_i = |c_i|^2$ and for arbitrary $\hat{\rho}_E \neq tr_S|\Psi\rangle\langle\Psi|$. Given $\hat{\rho}_E = \sum_\alpha \pi_\alpha |\alpha\rangle_E\langle\alpha|$, we decompose $|\Psi\rangle$ as:

$$|\Psi\rangle = \sum_{i,\alpha} c_i C_{i\alpha} |i\rangle_S |\alpha\rangle_E, \tag{30}$$

with the constraints:

$$\sum_i |c_i|^2 = 1 = \sum_\alpha \pi_\alpha, \sum_\alpha |C_{i\alpha}|^2 = 1, \forall i, \tag{31}$$

Then

$$\mathcal{Q}|\Psi\rangle\langle\Psi| = |\Psi\rangle\langle\Psi| - \sum_{i,\alpha} p_i \pi_\alpha |i\rangle_S \langle i| \otimes |\alpha\rangle_E \langle\alpha|. \tag{32}$$

We use ER:

$$|i\rangle_S |\alpha\rangle_E = \sum_{m,n} D^{i\alpha}_{mn} |m\rangle_{S'} |n\rangle_{E'} \tag{33}$$

with the constraints:

$$\sum_{m,n} D^{i\alpha}_{mn} D^{i'\alpha' *}_{mn} = \delta_{ii'}\delta_{\alpha\alpha'}. \tag{34}$$

With the use of eqs.(30) and (33), eq.(32) reads:

$$\sum_{m,m'n,n'} [\sum_{i,i',\alpha,\alpha'} c_i C_{i\alpha} c^*_{i'} C^*_{i'\alpha'} D^{i\alpha}_{mn} D^{i'\alpha' *}_{m'n'} - \sum_{i,\alpha} p_i \pi_\alpha D^{i\alpha}_{mn} D^{i\alpha\,*}_{m'n'}] |m\rangle_{S'} \langle m'| \otimes |n\rangle_{E'} \langle n'|. \tag{35}$$

After tracing out, $tr_{E'}$:

$$\sum_{m,m'} \{ \sum_{i,\alpha,n} \sum_{i',\alpha'} c_i C_{i\alpha} c^*_{i'} C^*_{i'\alpha'} D^{i\alpha}_{mn} D^{i'\alpha' *}_{m'n} - p_i \pi_\alpha D^{i\alpha}_{mn} D^{i\alpha *}_{m'n} \} |m\rangle_{S'} \langle m'| \tag{36}$$

Hence

$$tr_{E'} \mathcal{Q}|\Psi\rangle\langle\Psi| = 0 \Leftrightarrow \sum_{i,\alpha,n} [\sum_{i',\alpha'} c_i C_{i\alpha} c^*_{i'} C^*_{i'\alpha'} D^{i\alpha}_{mn} D^{i'\alpha' *}_{m'n} - p_i \pi_\alpha D^{i\alpha}_{mn} D^{i\alpha *}_{m'n}] = 0, \forall m, m'. \tag{37}$$

Introducing notation, $\Lambda^m_n \equiv \sum_{i,\alpha} c_i C_{i\alpha} D^{i\alpha}_{mn}$, one obtains:

$$tr_{E'} \mathcal{Q}|\Psi\rangle\langle\Psi| = 0 \Leftrightarrow A_{mm'} \equiv \sum_n [\Lambda^m_n \Lambda^{m' *}_n - \sum_{i,\alpha} p_i \pi_\alpha D^{i\alpha}_{mn} D^{i\alpha *}_{m'n}] = 0, \forall m, m'. \tag{38}$$

Notice:

$$\sum_m A_{mm} = 0. \tag{39}$$

which is equivalent to $tr \mathcal{Q}|\Psi\rangle\langle\Psi| = 0$, see eq.(32).
B) Mixed (e.g., non-entangled) state.

$$\hat{\rho} = \sum_i \lambda_i \hat{\rho}_{Si} \hat{\rho}_{Ei}, \quad \hat{\rho}_{Si} = \sum_m p_{im} |\chi_{im}\rangle_S \langle \chi_{im}|, \hat{\rho}_{Ei} = \sum_n \pi_{in} |\phi_{in}\rangle_E \langle \phi_{in}|, \tag{40}$$

In eq.(40), having in mind eq.(28), $tr_E \mathcal{Q}\hat{\rho} = 0$, while $tr_E \hat{\rho} = \sum_p \kappa_p |\varphi_p\rangle_S \langle \varphi_p|$, and $\hat{\rho}_E = \sum_q \omega_q |\psi_q\rangle_E \langle \psi_q| \neq tr_S \hat{\rho}$.

Constraints:

$$\sum_i \lambda_i = 1 = \sum_p \kappa_p = \sum_q \omega_q, \quad \sum_m p_{im} = 1 = \sum_n \pi_{in}, \forall i. \tag{41}$$

Now we make use of ER and, for comparison, we use the same basis $\{|a\rangle_{S'}|b\rangle_{E'}\}$

$$|\chi_{im}\rangle_S |\phi_{in}\rangle_E = \sum_{a,b} C_{ab}^{imn} |a\rangle_{S'} |b\rangle_{E'}, |\varphi_p\rangle_S |\psi_q\rangle_E = \sum_{a,b} D_{ab}^{pq} |a\rangle_{S'} |b\rangle_E. \tag{42}$$

Constraints:

$$\sum_{a,b} C_{ab}^{imn} C_{ab}^{im'n'*} = \delta_{mm'}\delta_{nn'}, \quad \sum_{a,b} D_{ab}^{pq} D_{ab}^{p'q'*} = \delta_{pp'}\delta_{qq'}. \tag{43}$$

So

$$\mathcal{Q}\hat{\rho} = \hat{\rho} - (tr_E \hat{\rho}) \otimes \hat{\rho}_E = \sum_{a,a',b,b'} \{ \sum_{i,m,n} \lambda_i p_{im} \pi_{in} C_{ab}^{imn} C_{a'b'}^{imn*} -$$

$$\sum_{p,q} \kappa_p \omega_q D_{ab}^{pq} D_{a'b'}^{pq*} \} |a\rangle_{S'} \langle a'| \otimes |b\rangle_{E'} \langle b'|. \tag{44}$$

Hence

$$tr_{E'} \mathcal{Q}\hat{\rho} = 0 \Leftrightarrow \Lambda_{aa'} \equiv \sum_{i,m,n,b} \lambda_i p_{im} \pi_{in} C_{ab}^{imn} C_{a'b}^{imn*}$$

$$- \sum_{p,q,b} \kappa_p \omega_q D_{ab}^{pq} D_{a'b}^{pq*} = 0, \forall a, a'. \tag{45}$$

Again, for $a = a'$:

$$\sum_a \Lambda_{aa} = 0, \tag{46}$$

as being equivalent with $tr\, \mathcal{Q}\hat{\rho} = 0$, see eq.(44).

Validity of eq.(13) assumes validity of eq.(38) for pure and of eq.(45) for mixed states. Both eq.(38) and eq.(45) represent the sets of simultaneously satisfied equations. We do not claim non-existence of the particular solutions to eq.(38) and/or to eq.(45), e.g., for the finite-dimensional systems. Nevertheless, we want to emphasize that the number of states they might refer to is apparently negligible compared to the number of states for which this is not the case. For instance, already for the fixed a and a', a small change e.g., in κs (while bearing eq.(41) in mind) undermines equality in eq.(45).

Quantum dynamics is continuous in time. Provided eq.(12) is fulfilled, validity of eq.(13) might refer *only* to a special set of the time instants. So we conclude: for the most part of the open S'-system's dynamics, eq.(13) is not fulfilled. By exchanging the roles of eq.(12) and eq.(13) in our analysis, we obtain the reverse conclusion, which completes the proof. Q.E.D.

Appendix B. Proof of Lemma 2

The commutation condition, $[\mathcal{P}, \mathcal{P}']\hat{\rho}(t) = 0, \forall t$. With the notation $\hat{\rho}_P(t) \equiv \mathcal{P}\hat{\rho}(t)$ and $\hat{\rho}_{P'}(t) \equiv \mathcal{P}'\hat{\rho}(t)$, the commutativity reads: $\mathcal{P}\hat{\rho}_{P'}(t) = \mathcal{P}'\hat{\rho}_P(t), \forall t$. Then $\mathcal{P}\hat{\rho}_{P'}(t) = tr_E\hat{\rho}_{P'}(t) \otimes \hat{\rho}_E = \hat{\rho}_S(t) \otimes \hat{\rho}_E$, while $\mathcal{P}'\hat{\rho}_P(t) = tr_{E'}\hat{\rho}_P(t)\otimes = \sigma_{S'}(t) \otimes \sigma_{E'}$. So, the commutativity requires the equality $\sigma_{S'}(t) \otimes \sigma_{E'} = \hat{\rho}_S(t) \otimes \hat{\rho}_E, \forall t$. However, quantum dynamics is continuous in time. Likewise in Proof of Lemma 1, quantum correlations relativity guarantees, that for the most of the time instants the equality will not be fulfilled. Q.E.D.

References

[1] Joos, E.; Zeh, H. D.; Kiefer, C.; Giulini, D.; Kupsch, J.; Stamatescu, I.-O. *Decoherence and the Appearance of a Classical World in Quantum Theory (2nd edition)*; Springer: Berlin, 2003.

[2] Zurek, W.H. *Rev. Mod. Phys.* 2003, 75, 715-775.

[3] Schlosshauer, M. *Rev. Mod. Phys.* 2005, 76, 1267-1305.

[4] Breuer, H.-P.; Petruccione, F. *The Theory of Open Quantum Systems*; Clarendon Press: Oxford, 2002.

[5] Rivas, Á.; Huelga, S. F. *Open Quantum Systems. An Introduction*; Springer: Berlin, 2011.

[6] Rivas, Á.; Huelga, S. F.; Plenio M. B. *Rep. Prog. Phys.* 2014, 77, 094001.

[7] Dugić, M.; Jeknić-Dugić, J. *Pramana J. Phys.* 2012, 79, 199-209.

[8] Lim, J.; Tame, M.; Yee, K. H.; Lee, J.-S.; Lee, J. *New J. Phys.* 2014, 16, 018001.

[9] Nakajima S. *Prog. Theor. Phys.* 1958, 20, 948-959.

[10] Zwanzig R. *J. Chem. Phys.* 1960, 33, 1338-1341.

[11] Dur, W.; Vidal, G. *Phys. Rev. A* 2000, 62, 062314.

[12] Eakins, J.; Jaroszkiewicz G. *J. Phys. A: Math. Gen.* 2003, 36, 517-526.

[13] Zanardi, P.; Lidar, D. A.; Lloyd, S. *Phys. Rev. Lett.* 2004, 92, 060402.

[14] Ciancio, E.; Giorda, P.; Zanardi P. *Phys. Lett. A* 2006, 354, 274-280.

[15] Dugić, M.; Jeknić, J. *Int. J. Theor. Phys.* 2006, 45, 2215-2225.

[16] Jeknić-Dugić, J.; Arsenijević, M.; Dugić, M. *Quantum Structures: A View of the Quantum World*; LAP Lambert Acad Publ: Saarbrücken, 2013.

[17] Dugić, M.; Arsenijević, M.; Jeknić-Dugić, J. *Sci. China Phys. Mech. Astron.* 2013, 56, 732-736.

[18] Brodutch, A.; Datta, D.; Modi, K.; Rivas, Á.; Rodriguez-Rosario, C. A. *Phys. Rev. A* 2013, 87, 042301.

[19] Gemmer, J.; Michel, M.; Mahler, G. *Quantum Thermodynamics, Lecture Notes in Physics*; Springer-Verlag: Berlin, 2004; Vol. 657.

[20] Arsenijević, M.; Jeknić-Dugić, J.; Dugić, M. (2013). A Limitation of the Nakajima-Zwanzig projection method. http://arxiv.org/abs/1301.1005.

[21] Caldeira, A. O.; Leggett, A. *Physica A* 1983, 121, 587.

[22] Stokes, A.; Kurcz, A.; Spiller, T. P.; Beige, A. *Phys. Rev. A* 2012, 85, 053805.

[23] Arsenijević, M.; Jeknić-Dugić, J.; Dugić, M. *Chin Phys B* 2013, 22, 020302.

[24] Arsenijević, M.; Jeknić-Dugić, J.; Dugić, M. (2013). Zero Discord for Markovian Bipartite Systems. http://arxiv.org/abs/1204.2789.

In: New Research on Quantum Entanglement ISBN: 978-1-63482-888-8
Editor: Lori Watson, pp. 117-123

Chapter 3

SCHEME OF GHOST IMAGING WITH TWO EPR STATES

D. L. Khokhlov*
Sumy State University (Retired), Ukraine

Abstract

A scheme of ghost imaging with two EPR states, AB and CD, is considered. The scheme involves both the type-one and type-two ghost imaging arrangements. Photons A and C are coupled by a 50/50 non-polarizing beam splitter. Photon A passes through an object. Photons B, C, D go freely. Photon D is captured by a bucket detector. Photon B is captured by a detector, scannable along the transverse direction. A ghost image of the object is observed in the joint-detection between the detectors, with the maximum visibility 50%.

1. Introduction

The concept of imaging came from classical optics. In a standard imaging setup, a lens of finite size is used to image the object onto an image plane, with the image plane and the object plane having a perfect point-to-point correspondence. In quantum optics, non-local two-photon states are used for ghost imaging, of type-one with entangled photons and type-two with thermal light [1]. The image is observed in the joint detection between two independent photodetectors, one

*E-mail address: dlkhokhl@rambler.ru

measures the scattered or transmitted light from the object, another measures light directly coming from the source at each point at the image plane. The point-to-point image-forming correlation between object plane and image plane is the result of two-photon interference, which involves the non-local superposition of two-photon amplitudes, corresponding to different yet indistinguishable alternative ways of triggering a joint-detection event. The first type-one ghost imaging experiment was demonstrated in 1995 [2] enlightened by the theoretical work of Klyshko [3]. The first near-field lensless type-two ghost imaging experiment was demonstrated in 2005-2006 [4].

A scheme with two EPR states was proposed in [5] for testing an interpretation of the measurement on EPR state [6]. In [7] the scheme was interpreted in terms of entanglement swapping. Unlike the standard scheme for entanglement swapping used for teleportation of a Bell state, the scheme [7] is intended for teleportation of the phase between the photons. In [8] it was shown that the scheme is based on the well-known quantum transformations: two-particle spatial correlation of entangled photons, teleportation, and two-photon interference of independent particles. In the present paper, a similar scheme is applied to ghost imaging.

2. The Physics behind Ghost Imaging

Consider the type-one ghost imaging with entangled photons, the lensless scheme. An EPR pair of orthogonally polarized signal and idler photons is fed into the setup. The signal and idler photons are sent in different directions by a polarizing beam splitter. The signal photon illuminates an aperture (object) and then is captured by a bucket detector D_1 which is placed at the focus of a short focal length collection lens. During the experiment D_1 is kept in a fixed position. The idler photon goes directly from the source to detector D_2, to which an optical fiber is coupled, the input tip of the fiber being scanned in the transverse plane. The outputs of both detectors are sent to a coincidence circuit for counting the joint-detection events of the pair. A ghost image of the object is observed in coincidences during the scanning of detector D_2, although the single detectors counting rates of D_1 and D_2 are both monitored to be constants in the object plane and in the image plane as well as in time.

The type-one ghost imaging experiment [2] is a realization of the EPR gedanken experiment [9]. The EPR correlation of the photons defined by the delta function of the global state is behind the type-one ghost imaging. The transverse spatial correlation of entangled photons is given by

$$\psi(\vec{\rho}_1, \vec{\rho}_2) \propto \int d\vec{k}_1 d\vec{k}_2 \delta(\vec{k}_1 + \vec{k}_2) e^{i\vec{k}_1\vec{\rho}_1} e^{i\vec{k}_2\vec{\rho}_2} \propto \delta(\vec{\rho}_1 - \vec{\rho}_2) \tag{1}$$

where $\vec{\rho}$ is the transverse coordinate of the detector, $\vec{k}$ is the transverse wave vector. The transverse wave vector of the first particle is changed by the object that gives rise to the change of the transverse wave vector of the second particle defined by the transverse momentum correlation $\delta(\vec{k}_1 + \vec{k}_2)$. It is worth noting that the transverse wave vector and correspondingly the transverse coordinate of the second particle taken individually remains unchanged in accordance with the peaceful coexistence, e.g., [10].

The transverse spatial correlation of the object and image is approximated as

$$G^{(2)}(\vec{\rho}_1, \vec{\rho}_2) \propto \delta(\vec{\rho}_1 - \vec{\rho}_2) \tag{2}$$

where $\vec{\rho}_1$, $\vec{\rho}_2$ are the transverse coordinate of the object and image planes respectively. The joint detection counting rate is given by

$$R_{12}(\vec{\rho}_2) \propto \int_{obj} d\vec{\rho}_1 |A(\vec{\rho}_1)|^2 \otimes \delta(\vec{\rho}_1 - \vec{\rho}_2) = |A(\vec{\rho}_2)|^2 \tag{3}$$

where A is the aperture function. This indicates an image of the object aperture function by means of the joint-detection events between distant detectors D_1 and D_2 while scanning D_2 in the image plane, with the maximum visibility 100%.

Consider the type-two ghost imaging with thermal light, the lensless scheme. The thermal radiation of a chaotic source is split in two spatial modes by a 50/50 non-polarizing beam splitter. The reflected light illuminates an aperture (object). The photons passing through the object are collected by a convex lens and counted by a bucket detector D_1. The transmitted light goes directly from the source to detector D_2, scannable in the transverse plane. The outputs of both detectors are sent to a coincidence circuit for counting the joint-detection between D_1 and D_2. A ghost image of the object is observed in the joint-detection between D_1 and D_2 during the scanning of detector D_2 in the transverse direction in the plane $z_1 = z_2$ where z_1 is the distance from the source to the object, z_2 is the distance from the source to the input tip of the fiber of detector D_2, being scanned in the transverse plane. The single detectors counting rates of D_1 and D_2 are both monitored to be constants in the object plane and in the image plane as well as in time.

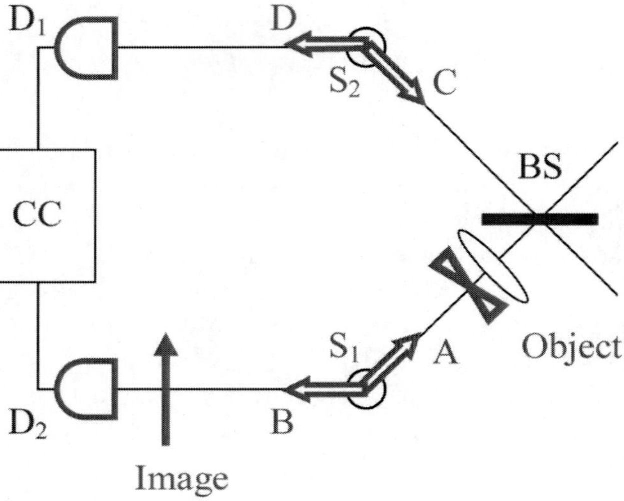

Figure 1. Scheme of experiment.

The partial correlation of thermal light is given by

$$G^{(2)}(\vec{\rho}_1, \vec{\rho}_2) \propto 1 + \delta(\vec{\rho}_1 - \vec{\rho}_2) \tag{4}$$

The joint detection counting rate is given by

$$R_{12}(\vec{\rho}_2) \propto \int_{obj} d\vec{\rho}_1 |A(\vec{\rho}_1)|^2 \otimes [1 + \delta(\vec{\rho}_1 - \vec{\rho}_2)] \propto const + |A(\vec{\rho}_2)|^2. \tag{5}$$

This indicates an image of the object aperture function by means of the joint-detection events between distant detectors D_1 and D_2 while scanning D_2 in the image plane, with the maximum visibility 50%.

3. Ghost Imaging with Two EPR States

Consider the scheme of ghost imaging with two EPR states, see Fig. 1. Sources S_1 and S_2 emit pairs of photons in the EPR states AB and CD respectively. Photons A and C are coupled by a 50/50 non-polarizing beam splitter. Photon A illuminates an aperture (object) and then is collected by a convex lens with a short focal length. The beam splitter is placed at the focus of the lens. Photons B, C, D go freely. Photon B is captured by a point-like detector D_2, scannable along the transverse direction. Photon D is captured by a bucket detector D_1 kept in a fixed position. The outputs of both detectors are sent to a coincidence circuit for counting the joint-detection events.

In the scheme, the transverse wave vector of photon A is changed by the aperture (object) that gives rise to the change of the transverse wave vector of photon B defined by the transverse momentum correlation $\delta(\vec{k}_A+\vec{k}_B)$. Accordingly, the delta function $\delta(\vec{\rho}_A - \vec{\rho}_B)$ defines the transverse spatial correlation of photons A and B. This corresponds to the type-one lensless ghost imaging, with the object plane at the place of the aperture and the image plane at the place of the fiber tip of detector D_2, scannable along the transverse direction. The distance from source S_1 to the object plane and the distance from source S_1 to the image plane are the same, with the magnification factor being equal unity.

The transverse spatial correlation of photons C and D are defined by the delta function $\delta(\vec{\rho}_C - \vec{\rho}_D)$. The transverse spatial correlation of independent photons A and C coupled at the beam splitter are defined by the function $1 + \delta(\vec{\rho}_A - \vec{\rho}_C)$. In effect, the transverse spatial correlation of photons A and D are defined by the function $[1 + \delta(\vec{\rho}_A - \vec{\rho}_C)]\delta(\vec{\rho}_C - \vec{\rho}_D)$. The correlation of photons A and D allows to observe the object through photon D captured by a bucket detector. Therefore, a ghost image of the object may be observed in the joint-detection events between detectors D_1 and D_2 counting photons D and B respectively, during the scanning of detector D_2, with the maximum visibility 50%. The ghost imaging requires the following distances to be of the same length: detector D_2 to source S_1, source S_1 to the beam splitter, the beam splitter to source S_2, source S_2 to detector D_1.

The 50/50 non-polarizing beam splitter is used in the type-two ghost imaging to split thermal light in two spatial modes. Here it serves to couple independent photons from two EPR states. When interchanging detectors D_1 and D_2, the arrangement corresponds to the type two ghost imaging whereas, instead of thermal light, two EPR states are used. Thus, the scheme under consideration involves both the type-one and type-two ghost imaging arrangements. The transformations between the photons in the scheme are those used in the type-one and type-two ghost imaging. Therefore, the scheme may be realized in the experiment.

4. Conclusion

We have considered a scheme of ghost imaging with two EPR states, AB and CD, with photons A and C being coupled by a 50/50 non-polarizing beam splitter. The arrangement for the state AB is typical for the lensless type-one ghost imaging, with photon A going through the object plane and photon B going through the image plane. In the type-one ghost imaging, photon A is captured by a bucket detector and photon B by a detector, scannable along the transverse direction. Due to the correlation of independent photons A and C coupled by the beam splitter as well as entangled photons C and D, one may observe the object through photon D captured by a bucket detector. A ghost image of the object is observed in the joint-detection between the detectors counting photons D and B, during the scanning of the detector counting photon B, with the maximum visibility 50%. The arrangement using the 50/50 non-polarizing beam splitter is typical for the type-two ghost imaging. Here, instead of thermal light used in the type-two ghost imaging, two EPR states are applied. Thus, the scheme involves both the type-one and type-two ghost imaging arrangements. The transformations between the photons in the scheme are those used in the type-one and type-two ghost imaging. Therefore, the scheme may be realized in the experiment.

References

[1] Y.H. Shih, *The physics of ghost imaging*, arXiv:0805.1166.

[2] T.B. Pittman, Y.H. Shih, D.V. Strekalov, and A.V. Sergienko, Optical imaging by means of two-photon quantum entanglement, *Phys. Rev. A* 52 (1995) R3429–3432.

[3] D.N. Klyshko, A simple method of preparing pure states of an optical field, of implementing the Einstein-Podolsky-Rosen experiment, and of demonstrating the complementarity principle, *Sov. Phys. Usp.* 31 (1988) 74–85; Combine EPR and two-slit experiments: Interference of advanced waves, *Phys. Lett. A* 132 (1988) 299–304.

[4] G. Scarcelli, V. Berardi, Y.H. Shih, Can two-photon correlation of chaotic light be considered as correlation of intensity fluctuations?, *Phys. Rev. Lett.* 96 (2006) 063602.

[5] D.L. Khokhlov, The scheme of the test of the measurement on EPR state, *Adv. Stud. Theor. Phys.*, 6 (2012) 791–795.

[6] D.L. Khokhlov, EPR scheme revisited, *Adv. Stud. Theor. Phys.*, 6 (2012) 601–605.

[7] D.L. Khokhlov, Proposal for the test with two EPR states, *Quantum Computers and Computing*, 12 (2012) 27–31.

[8] D.L. Khokhlov, A scheme of teleportation of the phase, *Nonlinear Optics Quantum Optics* 46(1) (2014) 19–27.

[9] A. Einstein, B. Podolsky, and N. Rosen, Can quantum mechanical description of physical reality be considered complete?, *Phys. Rev.* 47 (1935) 777–780.

[10] G.C. Ghirardi, A. Rimini, and T. Weber, A general argument against superluminal transmission through the quantum mechanical measurement process, *Lett. Nuovo Cimento* 27 (1980) 293–298.

INDEX

E

F

G

H

I

K

L

M

N

O

P

Q

R

S

T

U

V

W